BestMasters

Springer awards „BestMasters" to the best master's theses which have been com-pleted at renowned universities in Germany, Austria, and Switzerland.
The studies received highest marks and were recommended for publication by supervisors. They address current issues from various fields of research in natural sciences, psychology, technology, and economics.
The series addresses practitioners as well as scientists and, in particular, offers guid-ance for early stage researchers.

Marco Schröter

Dissipative Exciton Dynamics in Light-Harvesting Complexes

With a foreword by Prof. Dr. Oliver Kühn

Marco Schröter
Institute of Physics
University of Rostock
Germany

BestMasters
ISBN 978-3-658-09281-8   ISBN 978-3-658-09282-5 (eBook)
DOI 10.1007/978-3-658-09282-5

Library of Congress Control Number: 2015933635

Springer Spektrum

Printed on acid-free paper

Springer Spektrum is a brand of Springer Fachmedien Wiesbaden
Springer Fachmedien Wiesbaden is part of Springer Science+Business Media
(www.springer.com)

Foreword

„Learning from Nature" is a common theme in the natural sciences. The ingenious machinery of photosynthesis is particularly attractive as it has evolved to efficiently convert the sun light into chemically usable energy. Can we mimic natural photosynthesis to open the door to an unlimited energy source? Do we need to copy just structural elements into man made supramolecular architectures? How important is the interplay between structure and dynamics for the realization of function? Finally, did Nature restrict itself to the classical laws of Physics or did Quantum Mechanics provide some advantage? In fact it has been the last question which triggered most interest in the Physics community recently. Initiated by sophisticated nonlinear optical experiments, which revealed signatures of quantum coherent evolution at physiological temperatures, the first step of photosynthesis, i.e. the harvesting of sunlight, has been in the focus of broad research efforts.

A quantum dynamical description of a system as complex as a light-harvesting antenna protein provides quite a challenge. Models have to be developed, flexible enough to incorporate results from experiments as well as from atomistic simulations. Density matrix theory is the method of choice for dynamics simulations in condensed phases. With the recent development of efficient non-Markovian and non-perturbative approaches one is in the position to follow the quantum dynamics of the electronic excitations numerically exactly, which facilitates a test of models without invoking further approximations.

The present thesis is concerned with the dynamics and spectroscopy of electronic energy transfer in three model systems capturing different aspects of photosynthetic light harvesting. The simplest one, a molecular heterodimer, allows for a very detailed investigation of coherent oscillations in the dynamics. Here, it is possible to identify those quantum states, which are at the origin of these oscillations. In particular a scheme is devised, that enables discriminating between pure electronic and coupled electron-vibrational processes. This is of relevance since these types of coherences are affected differently by the protein environment, what might help to explain the observed long lasting oscillations. The second model mimics an energy funnel, a fundamental principle in photosynthesis. Special emphasis is devoted to the interplay between the energy level structure and relaxation and decoherence dynamics. The basic features of both models come together in the third application to the so-called Fenna-Matthews-Olson complex, which is part of the light-harvesting apparatus of green sulfur bacteria. This heterogeneous complex hosts different energy transfer pathways as well as coherent oscillations, which could be identified as being of electron-vibrational origin. Key to the success of this thesis has been the development of a numerical program package for the propagation of the density matrix and the extraction of linear and nonlinear spectroscopic signals.

The prospective reader of this thesis will benefit from the combination of mathematical derivations, numerical implementation, and specific applications to current problems in excitation energy transfer research in natural and artificial systems.

Prof. Dr. Oliver Kühn

Preface

Photosynthesis was studied intensively during the last decades by biologists, chemists, and physicists. Although the general process is well understood nowadays, the details, especially those concerning the effects leading to the high efficiency of the photosynthetic apparatus of plants, bacteria, and algae, require further investigations.

In the present work, an intermediate step in photosynthesis, that is the energy transfer from the light-absorbing antenna complexes to the photosynthetic reaction center is investigated from the perspective of theoretical physics. The concepts of dissipation theory and exciton dynamics are applied to a set of model aggregates to study various aspects, like transfer efficiency and spectral features, of the energy transfer in light-harvesting systems.

I like to thank the members of the molecular quantum dynamics group and the dynamics of molecular systems group at the University of Rostock, as well as the department of chemical physics at Lund University, who supported me during this project. Special thanks to Prof. Dr. Oliver Kühn, Prof. Dr. Tõnu Pullerits, B. Sc. Jan Schulze and Dr. Sergei Ivanov for various discussions. Finally, I would like to acknowledge the support of my family.

Marco Schröter

Institutional Profile

„Traditio et Innovatio" is the motto of the University of Rostock. While its history started as early as 1419, the beginnings of the Institute of Physics date back to the late 19th century. The Institute's galery of forefathers includes such famous names as Stern, Schottky, Hund, and Jordan. Nowadays, the interaction of radiation with matter is in the focus of research activities, which are linked by the Collaborative Research Center Sfb652 „Strong Correlations and Collective Effects in Radiation Fields". Further topics are, for instance, polymers, nanomaterials as well as surfaces and interfaces. From 2007 on the University's research profile has been shaped into four key directions, each associated with a department under the roof of an Interdisciplinary Faculty. The Department of „Life, Light & Matter" provides the frame for many research projects of the Institute of Physics, which cross the traditional borders of disciplines.

The research agenda of the „Molecular Quantum Dynamics" group headed by Prof. Oliver Kühn includes four topics. Non-reactive and reactive dynamics of nuclear degrees of freedom, Photophysics and Photochemistry of elementary processes, dynamics after X-ray core hole excitation, and Environmental Physics. The dynamics of nuclei is studied in the context of linear and nonlinear vibrational spectroscopy as a means to unravel the relation between molecular structure, dynamics, and function. Further, laser control theory is applied to manipulate dynamics such as to trigger, for instance, bond breaking. The arsenal of methods comprises those from quantum, semiclassical, and classical theory. Applications are concerned, for instance, with liquid water, ionic liquids or metal carbonyl compounds. Photoinduced processes in electronically excited molecular states are investigated with various electronic structure and dynamics methods.

Particular emphasis is put on systems relevant for photocatalysis and solar energy conversion. The present Master Thesis is an example for the research in the context of excitation energy transfer in man-made and natural light-harvesting antenna systems. With the advent of novel X-ray sources core-level spectroscopy has experienced a revival as a means to unravel, for instance, details of electronic structure and dynamics in situ. We focus, for instance, on transition metals in various environments, which are studied using first principles methods. Finally, an interdisciplinary effort is devoted to the introduction of atomistic simulation techniques into the field of soil science. Targets are pollutants and their interaction with soil components such as organic compounds or mineral surfaces.

In 2015 a new Physics building complex will be opened on the campus of the university. It will be situated next to the research building of the Department of „Life, Light & Matter", devoted to the science of „Complex Molecular Systems". We are looking forward to the new opportunities for interdisciplinary research, which will be provided by this excellent environment.

Contents

1. Introduction

Light conversion via photosynthesis by special bacteria, algae and green plants is the main source of the chemical energy used by higher organisms. The photosynthetic machinery of these organisms consists in general of light-harvesting antenna complexes and reaction centers, which are arranged such that they form an energy funnel directing the excitation energy flux towards the reaction center, Fig. 1.1. Sun energy absorbed by characteristic pigments in the antenna complexes, where an exciton is formed, is transferred to the reaction center. There, the excitation energy is converted, via charge separation processes like electron transfer, to a chemical redox potential, which drives chemical reactions. The overall quantum efficiency of the light-harvesting process is very high ($> 95\%$) due to a very efficient energy transfer in the photosynthetic complexes [1].

The mechanism of excitation energy transfer in photosynthetic complexes has been studied extensively for decades using Förster or Frenkel exciton theory [2–15]. It has caught renewed interest due to the observation of long lasting coherent oscillations in various light-harvesting complexes via electronic coherent 2D-spectroscopy experiments [16]. In particular the question, whether the observed oscillations in the 2D-spectra of light-harvesting complexes are of electronic or vibrational origin and their connection to the excellent energy transfer properties, remains controversial [17–30].

The protein environment and intra-molecular vibrations, which cause, e.g., static and dynamic disorder of the excitation energies and/or the inter-molecular couplings, create energy funnels within the aggregates. This leads to a spatially directed energy transfer connected with an energy relaxation towards the reaction center. Recent investigations concerning small aggregates, i.e. dimers, unveiled

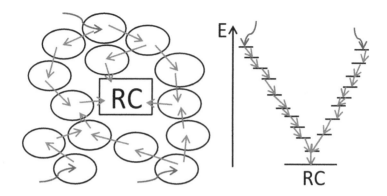

Figure 1.1.: Schematic picture of the photosynthetic energy transfer process. The outer antenna complexes absorb the photon energy (red wavy arrows), which is transferred via an exciton transfer mechanism (pathways indicated by blue straight arrows) to the reaction center (RC).

that in particular vibronic couplings and resonances between electronic and vibronic levels play a distinct role for the energy transfer [22, 24, 26, 27].

The influence of the environment can be included into theoretical models in the spirit of a system-bath ansatz, e.g., via a perturbative treatment like in the Redfield approach [31]. Such approaches, although applicable to many systems, are limited to weak interactions and might not cover all important effects of the system-bath interaction. Exact methods like path integral approaches, however, are limited to rather small systems [32, 33]. The hierarchy equations of motion (HEOM) formalism, initially proposed by Kubo and Tanimura [34] in 1989, provides an in principle exact treatment of the system-bath interaction and is by virtue of recent improvements efficient enough to investigate energy transfer processes in systems like light-harvesting complexes [35–37].

In the following section the theoretical concepts, which are used to study the exciton dynamics in molecular aggregates embedded in an environment, will be introduced. First, the system-bath Hamiltonian ansatz will be summarized. Second, the quantum master equation (QME) and HEOM formalisms, which provide propagation schemes for the reduced density matrix describing the system of interest, will be discussed. Both formalisms treat the system-bath interaction by means of the bath correlation function, which is usually calculated via the so-called spectral density, that will be introduced in Sec. 2.3. A brief summary of the response function formalism in Sec. 2.4. It provides the connection between the evolution of the reduced density matrix and the linear and nonlinear spectra of the investigated systems is given

The outlined theoretical concepts will be used to investigate the dissipative exciton dynamics in dependence on the system and bath properties of different model aggregates. First, a molecular dimer system will be characterized in terms of population dynamics, linear absorption, and 2D-spectra. In particular, the oscillatory features of the population dynamics, which might be connected to oscillations in 2D-spectra, will be analysed. Second, an aggregate, which represents an energy funnel as a generic model for light-harvesting antennae, will be studied. Here, the focus is on the influence of the bath properties on the excitation energy pathways within the aggregate. Finally, a particular example for a light-harvesting complex, the Fenna-Matthews-Olson (FMO) complex, will be treated, utilizing experimental data for the spectral density.

All aforementioned examples have been studied with the HEOM approach. The numerical implementation of this method is provided by the Rostock HEOM package, which was developed during the present master project. Detailed information about the package is given in Appendix A. The code is available upon request.

2. Dissipative quantum dynamics

Complex biological systems, like the photosynthetic units of bacteria or plants, usually consist of several pigment molecules, which form aggregates, embedded in a protein environment. Within a quantum mechanical treatment it is practically impossible to take all degrees of freedom (DOFs) of such large systems explicitly into account. Note that there exist highly efficient methods, e.g., the multiconfiguration time-dependent Hartree method (MCTDH) [38], to treat the quantum dynamics of isolated systems, e.g., the intra-molecular dynamics of the aforementioned pigment molecules. The recently developed multilayer expansion of MCTDH [39, 40] facilitates the treatment of systems with a few thousand DOFs, which is sufficient to handle small systems within an environment. This approach requires a discretization of the spectral density, cf. Sec. 2.3, and is therefore restricted in its applicability. An application of the MCTDH approach to the dissipative quantum dynamics of a light-harvesting system is given in Ref. [41]. However, it is preferable to reduce the dimensionality of the problem considering only few relevant DOFs, which provide insights into the physical processes, explicitly and to treat all other DOFs as a heat bath interacting with the system DOFs via an energy exchange. According to this separation the Hamiltonian of the full system is given by

$$H = H_s + H_b + H_{s\text{-}b}, \qquad (2.1)$$

where $H_{s/b}$ denotes the system/bath Hamiltonian and $H_{s\text{-}b}$ the interaction between the system and the bath. The separation allows one to solve the quantum mechanical equations of motion for the relevant system DOFs considering the influence of the bath DOFs exactly or approximately.

In the following section the Hamiltonian describing aggregates of pigment molecules embedded in a protein environment will be discussed. Further, two different sets of equations of motion for the reduced density matrix, namely the QME and the in principle exact HEOM will be introduced in Sec. 2.2. Both incorporate the influence of the bath via a spectral density function. The latter and its connection to the Hamiltonian, as well as some models for the spectral density, will be discussed in Sec. 2.3. Finally, the optical response function formalism, which connects the time evolution of the density matrix to the optical spectra of the system, will be briefly introduced in Sec. 2.4.

2.1. Hamiltonian

The system Hamiltonian, H_s, of an aggregate consisting of N_{agg} monomers is given by the Frenkel exciton Hamiltonian. It is based on the assumption that the interacting monomers, which form the aggregate, retain their chemical identity. According to this assumption the monomeric adiabatic states $|m_a\rangle$ can be used to construct diabatic aggregate states, which are coupled due to the inter-monomeric Coulomb coupling (CC). Here, $a = g, e, \ldots$ labels the adiabatic monomeric electronic state and $m = 1, \ldots, N_{agg}$ denotes the monomer. The aggregate states can be classified according to the number of electronic excitations within the aggregate. Considering only one excited state, $a = e$, per monomer, the aggregate ground state $|g\rangle$ and the one-exciton states $|e_m\rangle$ are given by

$$|g\rangle = \prod_m |m_g\rangle \tag{2.2}$$

and

$$|e_m\rangle = |m_e\rangle \prod_{n \neq m} |n_g\rangle. \tag{2.3}$$

Higher-order excitations, such as the excitation of higher monomeric adiabatic states, e.g., the S_n states of pigment molecules, or multiple excitations within the aggregate, can be incorporated analogously into the description. Including only the ground and one-exciton states, the Frenkel Hamiltonian is given by

$$H_{\text{agg}} = H(g)|g\rangle\langle g| + \sum_m H_m(e_m)|e_m\rangle\langle e_m|$$
$$+ \sum_{m,n} J_{mn}(e_m, e_n)|e_m\rangle\langle e_n| + H_{\text{int}}$$
$$= H_{\text{s}}^{(0)} + H_{\text{s}}^{(1)} + H_{\text{int}} \tag{2.4}$$

with the zero- and one-excitation parts

$$H_{\text{s}}^{(0)} = H(g)|g\rangle\langle g| \tag{2.5}$$

and

$$H_{\text{s}}^{(1)} = \sum_m H_m(e_m)|e_m\rangle\langle e_m| + \sum_{m,n} J_{mn}(e_m, e_n)|e_m\rangle\langle e_n|, \tag{2.6}$$

respectively. The coupling elements $J_{mn}(e_m, e_n)$ represent the interaction between different one-exciton states and H_{int} describes, e.g., the interaction of the system with external fields.

The diagonal terms of the aggregate Hamiltonian are, according to the definition of the exciton states, Eq. (2.2) and Eq. (2.3), given by

$$H(g) = \sum_m H(m_g) \tag{2.7}$$

and

$$H_m(e_m) = H(m_e) + \sum_{n \neq m} H(n_g). \tag{2.8}$$

Here, $H(m_g)$ and $H(m_e)$ denote the Hamiltonian of the adiabatic monomeric ground and excited state of monomer m, respectively. The diabatic aggregate states depend on the set $\{R\} = (R_1, \ldots, R_{N_{\text{agg}}})$ of all intra-molecular nuclear coordinates, i.e. $|g\rangle = |g(\{R\})\rangle$ and $|e_m\rangle = |e_m(\{R\})\rangle$. This is due to the parametric dependence of the

monomeric adiabatic states on the monomeric set of intra-molecular
nuclear coordinates $R_m = (\vec{R}_{m,1}, \vec{R}_{m,2}, \ldots)$, i.e. $|m_g\rangle = |m_g(R_m)\rangle$
and $|m_e\rangle = |m_e(R_m)\rangle$. Note that the intra-molecular vibrations are
separated from the inter-molecular vibrations. The latter, as well as
the vibrations of the protein environment and their influence on the
system, can be conveniently described by the bath and the system-
bath Hamiltonian. The dependence of the diabatic aggregate states
on the intra-molecular nuclear DOFs can be treated implicitly by
including the intra-molecular vibrations into the heat bath as well,
or explicitly, e.g., by the shifted oscillator model [31]. In the former
case the on-diagonal terms $H(g)$ and $H_m(e_m)$ are given by the bare
electronic state energies E_g and E_{e_m}, respectively. However, in the
latter case the electronic state energies need to be supplemented by
a contribution representing the change of the potential energy with
respect to $\{R\}$ and the corresponding kinetic energy contribution.
The corresponding on-site elements of the system Hamiltonian are
given by

$$H(g, \{R\}) = E_g + U_g(\{R\}) + T_g(\{R\}) \tag{2.9}$$

and

$$H_m(e_m, \{R\}) = E_{e_m} + U_{e_m}(\{R\}) + T_{e_m}(\{R\}). \tag{2.10}$$

A second order Taylor expansion of the potential energy $U(\{R\})$
around its minimum with respect to $\{R\}$ in combination with a
normal mode transformation yields the frequently used harmonic
oscillator model, i.e.

$$H(g, \{q\}) = E_g + \frac{1}{2} \sum_m \sum_\xi \left(p_{\xi,m}^{(g)\,2} + \omega_{\xi,m}^{(g)\,2} q_{\xi,m}^{(g)\,2} \right) \tag{2.11}$$

and

$$H_m(e_m, \{q\}) = E_{e_m} + \frac{1}{2} \sum_{n \neq m} \sum_\xi \left(p_{\xi,n}^{(g)\,2} + \omega_{\xi,n}^{(g)\,2} q_{\xi,n}^{(g)\,2} \right)$$
$$+ \frac{1}{2} \sum_\xi \left(p_{\xi,m}^{(e)\,2} + \omega_{\xi,m}^{(e)\,2} q_{\xi,m}^{(e)\,2} \right). \tag{2.12}$$

Here, $p_{\xi,m}^{(a)}$ denotes the momentum and $\omega_{\xi,m}^{(a)}$ the harmonic frequency associated with the ξth mode of monomer m, being in the electronic state a, in normal mode space $\{q\}$. Note that the normal coordinates are mass-weighted and that $\hbar = 1$ here and in the following. It is convenient to define the normal modes in the corresponding monomeric ground states $|m_g\rangle$ and project all vibrational properties of the aggregate states $|g\rangle$ and $|e_m\rangle$ onto this basis. Assuming additionally that there are no changes in the curvature of the oscillator potentials yields the shifted oscillator model. In the limit of the aforementioned approximations the on-site Hamiltonian is given by

$$H(g, \{q\}) = E_g + \frac{1}{2} \sum_m \sum_\xi \left(p_{\xi,m}^2 + \omega_{\xi,m}^2 q_{\xi,m}^2 \right) \qquad (2.13)$$

for the ground state and by

$$H_m(e_m, \{q\}) = E_{e_m} + \frac{1}{2} \sum_\xi \left(p_{\xi,n}^2 + \omega_{\xi,n}^2 q_{\xi,n}^2 \right)$$

$$+ \frac{1}{2} \sum_\xi \sum_{n \neq m} \left(p_{\xi,m}^2 + \omega_{\xi,m}^2 (q_{\xi,m} - d_{\xi,m})^2 \right) \qquad (2.14)$$

for the excited states, respectively. Here, $d_{\xi,m}$ represents the shift of the excited state oscillator along $q_{\xi,m}$ with respect to the ground state potential energy surface. Introducing intrinsic harmonic oscillator variables, i.e.

$$\tilde{p}_{\xi,m} = \sqrt{\frac{1}{\omega_{\xi,m}}} p_{\xi,m} \qquad (2.15)$$

and

$$\tilde{q}_{\xi,m} = \sqrt{\omega_{\xi,m}} q_{\xi,m}, \qquad (2.16)$$

yields

$$H(g, \{\tilde{q}\}) = E_g + \sum_m \sum_\xi \frac{\omega_{\xi,m}}{2} \left(\tilde{p}_{\xi,m}^2 + \tilde{q}_{\xi,m}^2 \right) \qquad (2.17)$$

and

$$H_m(e_m, \{\tilde{q}\}) = E_{e_m} + \sum_{n \neq m} \sum_{\xi} \frac{\omega_{\xi,n}}{2} \left(\tilde{p}_{\xi,n}^2 + \tilde{q}_{\xi,n}^2 \right)$$
$$+ \sum_{\xi} \frac{\omega_{\xi,m}}{2} \left(\tilde{p}_{\xi,m}^2 + (\tilde{q}_{\xi,m} - \tilde{d}_{\xi,m})^2 \right). \qquad (2.18)$$

The dimensionless shift $\tilde{d}_{\xi,m}$ can be expressed by the Huang-Rhys factor $S_{\xi,m} \equiv \tilde{d}_{\xi,m}^2/2$, which is often used to characterize the coupling between electronic and vibrational DOFs. Expanding Eq. (2.18) using the definition of $S_{\xi,m}$ yields

$$H_m(e_m, \{\tilde{q}\}) = E_{e_m} + \sum_{\xi} \sum_{n} \frac{\omega_{\xi,n}}{2} \left(\tilde{p}_{\xi,n}^2 + \tilde{q}_{\xi,n}^2 \right)$$
$$- \sum_{\xi} \omega_{\xi,m} \sqrt{2 S_{\xi,m}} \tilde{q}_{\xi,m} + \sum_{\xi} \omega_{\xi,m} S_{\xi,m}. \qquad (2.19)$$

Note that the last sum represents the reorganization energy associated with the relaxation within the excited state after a vertical transition from the ground to the excited state and is thus related to the Stokes shift in linear spectroscopy.

The interaction between configurations of the aggregate, where the excitation is located on different sites of the aggregate, i.e. between different one-exciton states, is in general (in terms of the monomeric adiabatic states) given by the Coulomb integral

$$J_{mn}(m_a n_b, n_c m_d) = \int d\vec{r} d\vec{r}' \frac{\mathfrak{N}_{m_a,m_d}(\vec{r}) \mathfrak{N}_{n_b,n_c}(\vec{r}')}{|\vec{r} - \vec{r}'|}. \qquad (2.20)$$

Here, $\vec{r}$ denotes an electronic coordinate and $\mathfrak{N}$ stands for the generalised molecular charge density [31] depending on the adiabatic monomeric states. Note that within the present model only elements, which describe a simultaneous excitation of one and de-excitation of another monomer, are considered. Thus, the inter-monomeric interactions between two sites m and n in Eq. (2.4) are represented by the coupling elements $J_{mn}(e_m, e_n)$, which are given by

$$J_{mn}(e_m, e_n) = \int \mathrm{d}\vec{r}\mathrm{d}\vec{r}' \frac{\tilde{\mathfrak{N}}_{e_m}(\vec{r})\tilde{\mathfrak{N}}_{e_n}(\vec{r}')}{|\vec{r} - \vec{r}'|}, \qquad (2.21)$$

where $\tilde{\mathfrak{N}}$ denotes the generalized charge density associated with the corresponding one-exciton state.

The Coulomb integral, Eq. (2.21), can be evaluated, e.g., by employing a transition dipole approximation, assuming that the separation of the monomers is large compared to the spatial extension of the transition density [31]. However, more elaborate methods, such as the transition density cube [5], the transition charge from electrostatic potential method [42], or time-dependent tight-binding-based density functional theory [43], facilitate the calculation of more accurate Coulomb coupling matrix elements.

The diabatic one-exciton aggregate Hamiltonian, Eq. (2.6), can be transformed to an adiabatic representation via diagonalization of its potential energy part. This leads to a set of adiabatic (delocalized) states $|\alpha\rangle$ which are connected to the diabatic states via the components $c_{\alpha,m}$ of the eigenvectors of the diabatic Hamiltonian, i.e.

$$|\alpha\rangle = \sum_m c_{\alpha,m} |e_m\rangle. \qquad (2.22)$$

For example, for an electronic dimer system the energy of the adiabatic excited states is given by [31]

$$E_{1,2} = \frac{E_{e_1} + E_{e_2}}{2} \mp \frac{1}{2}\sqrt{(E_{e_1} - E_{e_2})^2 + 4J_{12}^2}. \qquad (2.23)$$

Note that the adiabatic levels are numbered according to their energy in increasing order, i.e. the level with the lowest energy is labelled with 1, the second lowest with 2 and so forth. The aggregate ground state $|g\rangle$ is the same in both representations as there exists no coupling to the one-exciton states.

The last term in Eq. (2.4) describes interactions with external fields. In dipole approximation H_{int} is given by

$$H_{\text{int}} = -d\mathfrak{E}(t), \tag{2.24}$$

for the interaction with an external laser field assuming that the field $\vec{\mathfrak{E}}(t)$ and the transition dipole moment $\vec{d}$ exhibit a constant orientation with respect to each other, i.e.

$$\vec{d} \cdot \vec{\mathfrak{E}}(t) = c \cdot |d| \, |\mathfrak{E}(t)| \,. \tag{2.25}$$

Here, c denotes an orientation factor, which is assumed to be included in the definition of d in Eq. (2.24). The aggregate dipole operator with the site-dependent dipole transition strength $d_{e_m,g}$ is given by

$$d = \sum_m d_{e_m,g} |e_m\rangle\langle g| + \text{h.c.}. \tag{2.26}$$

Under certain conditions all kinds of environments, even those that contain anharmonic effects, can be described by an effective harmonic bath [32]. However, the specific form of the bath as well as of the system-bath Hamiltonian depends on the underlying model. First, consider the case that only the DOFs of the environment contribute to the bath. This is the case if the intra-molecular nuclear DOFs can be neglected, e.g., if their coupling to the electronic DOFs is negligible or if they are included in the system explicitly. The Hamiltonian of the environmental DOFs, $H_{\text{b}}^{(\text{I})}$, is given by

$$H_{\text{b}}^{(\text{I})} = \sum_i \frac{\omega_i}{2} \left(\tilde{p}_i^2 + \tilde{x}_i^2 \right). \tag{2.27}$$

Here, ω_i denotes the harmonic frequency and $\tilde{x}_i$ the corresponding bath coordinate. If the intra-molecular vibrations contribute to the bath, it is convenient to separate the bath Hamiltonian into an intra-molecular and an environmental part, i.e.

$$H_{\text{b}}^{(\text{II})} = H_{\text{b}}^{(\text{int})} + H_{\text{b}}^{(\text{env})}. \tag{2.28}$$

The intra-molecular part, $H_{\mathrm{b}}^{(\mathrm{int})}$, is given by, cf. Eq. (2.19),

$$H_{\mathrm{b}}^{(\mathrm{int})} = \sum_m \sum_\xi \frac{\omega_{\xi,m}}{2} \left(\tilde{p}_{\xi,m}^2 + \tilde{q}_{\xi,m}^2 \right) \tag{2.29}$$

and the environmental part by Eq. (2.27). Thus, the full bath Hamiltonian reads

$$H_{\mathrm{b}}^{(\mathrm{II})} = \sum_m \sum_\xi \frac{\omega_{\xi,m}}{2} \left(\tilde{p}_{\xi,m}^2 + \tilde{q}_{\xi,m}^2 \right) + \sum_i \frac{\omega_i}{2} \left(\tilde{p}_i^2 + \tilde{x}_i^2 \right). \tag{2.30}$$

The system-bath interaction Hamiltonian, $H_{\mathrm{s\text{-}b}}$, depends in general on the system and bath operators. However, it can be decomposed into a product of arbitrary functions of system and bath operators, i.e.

$$H_{\mathrm{s\text{-}b}} = \Phi_{\mathrm{b}}(X) K_{\mathrm{s}}(Q), \tag{2.31}$$

with no loss of generality. A linear Taylor expansion of the bath function $\Phi_{\mathrm{b}}(X)$ and of the system function $K_{\mathrm{s}}(Q)$ yields the widely used Caldeira-Leggett model [44] for the system-bath Hamiltonian

$$H_{\mathrm{s\text{-}b}} = \sum_j \sum_\zeta c_{j,\zeta} X_j Q_\zeta. \tag{2.32}$$

Here, X_j denotes an harmonic oscillator position operator and Q_ζ an operator representing a part of the system. Note that the linear expansion with respect to the system operators restricts the description to energy gap fluctuations in the system. An expansion of $K_{\mathrm{s}}(Q)$ up to higher orders of Q would take additional effects into account, e.g., a modulation of the curvature of the potential energy surface [45].

The general form of the Caldeira-Leggett system-bath Hamiltonian, Eq. (2.32), needs to be adjusted according to the bath and system-bath models (SBMs). Note that the same bath configuration leads in combination with different SBMs to a different dynamics of the system.

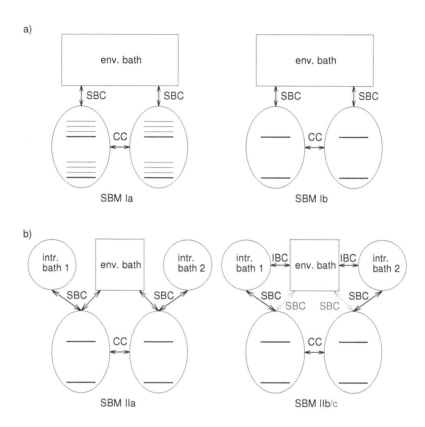

Figure 2.1.: Schematic view of the system-bath models (SBM) for a dimer system. **(a)** In SBM Ia/b the DOFs of the two monomers are coupled to the environmental DOFs by the system-bath coupling (SBC) and to each other via the Coulomb coupling (CC). SBM Ia and b differ in the description of the system. In the former the intra-molecular vibrations are included in the system, whereas they are neglected in the latter. **(b)** The intra-molecular vibrations are treated as separated bath. In contrast to SBM IIa, where the environmental bath is not coupled to the intra-molecular one, there exists an inter-bath coupling (IBC) in SBM IIb/c.

The system-bath Hamiltonian for bath model I, see Eq. (2.27), assuming that the intra-molecular DOFs are included in the system explicitly, is given by

$$H_{\text{s-b}}^{(\text{Ia})} = \sum_i \sum_{\xi,m} c_{i,\xi m} \tilde{x}_i \tilde{q}_{\xi,m} |e_m\rangle\langle e_m|. \tag{2.33}$$

Here, the index ζ in Eq. (2.32) is associated with the combination of mode and monomer index ($\zeta = \xi, m$), the general system coordinate Q_ζ equals $\tilde{q}_{\xi,m}|e_m\rangle\langle e_m|$ and the general bath coordinates X_j are identical with the coordinates of the environmental DOFs, i.e. $X_j = \tilde{x}_i$. If the intra-molecular DOFs can be neglected the expression simplifies to

$$H_{\text{s-b}}^{(\text{Ib})} = \sum_i \sum_m c_{i,m} \tilde{x}_i |e_m\rangle\langle e_m|. \tag{2.34}$$

Both scenarios are depicted in Fig. 2.1 panel a. Bath model II, see Eq. (2.30) leads to different coupling schemes (SBM IIa-c, Fig. 2.1 panel b). However, the discussion will be restricted to SBM IIa and b in the following as the system-bath Hamiltonian for SBM IIc can be constructed analogously. SBM IIa refers to the situation where the system couples separately to the baths of intra-molecular DOFs and to the bath(s) of environmental DOFs, but that the former are not coupled to the latter. The corresponding system-bath Hamiltonian is given by

$$H_{\text{s-b}}^{(\text{IIa})} = \sum_m \sum_\xi \omega_{\xi,m} \sqrt{2 S_{\xi,m}} \tilde{q}_{\xi,m} |e_m\rangle\langle e_m|$$
$$+ \sum_m \sum_i c_{i,m} \tilde{x}_i |e_m\rangle\langle e_m|. \tag{2.35}$$

In contrast, SBM IIb refers to the situation that the system couples to primary baths of intra-molecular DOFs which couple to secondary bath(s) of environmental DOFs. The system-bath Hamiltonian for SBM IIb is given by

$$H_{\text{s-b}}^{(\text{IIb})} = \sum_m \sum_\xi \omega_{\xi,m} \sqrt{2 S_{\xi,m}} \tilde{q}_{\xi,m} |e_m\rangle\langle e_m|, \tag{2.36}$$

but note that the bath Hamiltonian needs to be supplemented by
an interaction term $H_{\text{bp-bs}}$ between the primary and the secondary
bath [46], which is given by

$$H_{\text{bp-bs}}^{(\text{IIb})} = \sum_m \sum_\xi \sum_i c_{i,\xi m} \tilde{q}_{\xi,m} \tilde{x}_i. \qquad (2.37)$$

Note further that the introduced system-bath models account only
for fluctuations of the diagonal elements of the system Hamiltonian.
In principle also the off-diagonal Coulomb coupling matrix elements
could be affected by the system-bath interaction [23].

2.2. Equations of motion for the reduced density matrix

The system and bath DOFs are still coupled due to system-bath
interaction term $H_{\text{s-b}}$ in Eq. (2.1). However, it is possible to construct
equations of motion for the so-called reduced density matrix $\rho_s(t) =$
$\text{tr}_{\text{bath}} \{\rho(t)\}$, which describes the influence of the bath on the system
dynamics implicitly. A perturbative treatment of the system-bath
interaction in combination with the Markov approximation leads to the
QME which is widely used to investigate the energy transfer in light-
harvesting complexes [11, 21, 22, 31, 47, 48]. QME approaches are
only suitable for weak system-bath interactions as memory effects of
the bath and higher-order interactions are neglected [49]. In contrast,
the HEOM formalism, whose development was initialized by Tanimura
and Kubo in 1989 [34], provides an in principle exact treatment of
the system-bath interaction.

2.2.1. Quantum master equation

The time evolution of the full density matrix

$$\rho(t) = U(t,t_0)\rho(t_0)U^\dagger(t,t_0) = \mathcal{U}(t,t_0)\rho(t_0) \qquad (2.38)$$

is determined by the full system Hamiltonian H via the time evolution operator

$$U(t, t_0) = e^{-iH(t-t_0)} \tag{2.39}$$

or its Liouville space equivalent $\mathcal{U}(t, t_0)$. According to the separation of the Hamiltonian, Eq. (2.1), the formal equation of motion of the full density matrix is given by the Liouville-von Neumann equation

$$\frac{\partial}{\partial t}\rho(t) = -i\,[H, \rho(t)]$$
$$= -i\,[H_{\mathrm{s}}, \rho(t)] - i\,[H_{\mathrm{b}}, \rho(t)] - i\,[H_{\mathrm{s\text{-}b}}, \rho(t)]. \tag{2.40}$$

In the interaction representation the corresponding equation is given by

$$\frac{\partial}{\partial t}\rho^{(\mathrm{I})}(t) = -i\left[H_{\mathrm{s-b}}(t), \rho^{(\mathrm{I})}(t)\right], \tag{2.41}$$

where the initial full density matrix in the interaction representation is defined as

$$\rho^{(\mathrm{I})}(t_0) = U_0^\dagger(t, t_0)\rho(t_0)U_0(t, t_0). \tag{2.42}$$

The time evolution operator $U_0(t, t_0)$ is defined in analogy to $U(t, t_0)$, Eq. (2.39), replacing H by the Hamiltonian of the non-interacting subsystems $H_{\mathrm{s}} + H_{\mathrm{b}}$. Inserting the formal integration of Eq. (2.41) into the right-hand side of Eq. (2.41), that is essentially applying second-order perturbation theory with respect to $H_{\mathrm{s-b}}$, yields

$$\frac{\partial}{\partial t}\rho^{(\mathrm{I})}(t) = -i\left[H_{\mathrm{s-b}}(t), \rho^{(\mathrm{I})}(t_0)\right]$$
$$- \int_{t_0}^{t} dt'\left[H_{\mathrm{s-b}}(t), \left[H_{\mathrm{s-b}}(t'), \rho^{(\mathrm{I})}(t')\right]\right]. \tag{2.43}$$

Assuming further uncorrelated initial conditions for the density matrix, i.e.

$$\rho^{(\mathrm{I})}(t_0) = \rho_{\mathrm{s}}(t_0)\rho_{\mathrm{b}}(t_0), \tag{2.44}$$

where $\rho_{s/b}(t_0)$ denotes the equilibrium density matrix of the system/bath, and that the bath stays in equilibrium independently of the amount of energy transferred to it by the system, which is fulfilled if the number of bath DOFs is sufficiently large, the density matrix at an arbitrary time is given by

$$\rho^{(I)}(t) = \rho_s^{(I)}(t)\rho_b(t_0).\tag{2.45}$$

This leads to

$$\frac{\partial}{\partial t}\rho_s^{(I)}(t) = -\operatorname{tr}_{\text{bath}}\left\{i\left[H_{s-b}(t), \rho_s(t_0)\rho_b(t_0)\right]\right\}$$

$$-\int_{t_0}^{t} dt'\operatorname{tr}_{\text{bath}}\left\{\left[H_{s-b}(t), \left[H_{s-b}(t'), \rho_s^{(I)}(t')\rho_b(t_0)\right]\right]\right\}.$$

$$\tag{2.46}$$

The integral on the right-hand side of Eq. (2.46) depends on all former time points due to $\rho_s^{(I)}(t')$. Therefore, the reduced density matrix memorizes its own evolution. Expanding the commutators using Eq. (2.31) and taking into account that the system and the bath operators commute leads to

$$\frac{\partial}{\partial t}\rho_s^{(I)}(t)$$
$$= -iK_s(Q(t))\rho_s(t_0)\operatorname{tr}_{\text{bath}}\left\{\Phi_b(X(t))\rho_b(t_0)\right\}$$
$$+ i\rho_s(t_0)K_s(Q(t))\operatorname{tr}_{\text{bath}}\left\{\Phi_b(X(t))\rho_b(t_0)\right\}$$
$$- \Im(K_s(Q(t)), K_s(Q(t')), \rho_s^{(I)}(t'), \Phi_b(X(t)), \Phi_b(X(t')))$$
$$+ \Im(K_s(Q(t)), \rho_s^{(I)}(t'), K_s(Q(t')), \Phi_b(X(t')), \Phi_b(X(t)))$$
$$+ \Im(K_s(Q(t')), \rho_s^{(I)}(t'), K_s(Q(t)), \Phi_b(X(t)), \Phi_b(X(t')))$$
$$- \Im(K_s(Q(t')), K_s(Q(t)), \rho_s^{(I)}(t'), \Phi_b(X(t')), \Phi_b(X(t))),\tag{2.47}$$

where $\Im(A, B, C, D, E)$ is defined as

$$\Im(A, B, C, D, E) = \int_{t_0}^{t} dt'\, A \cdot B \cdot C \operatorname{tr}_{\text{bath}}\left\{D \cdot E \cdot \rho_b(t_0)\right\}.\tag{2.48}$$

Note that the trace over $\Phi_b(X(t))\rho_b(t_0)$ is actually the expectation value of $\Phi_b(X(t))$, which is zero by definition for a harmonic bath, cf. Eq. (2.27). Thus, the first two terms of Eq. (2.47) vanish in this case. Rewriting Eq. (2.47) and introducing the bath correlation function

$$C(t - t') = \text{tr}_{\text{bath}} \left\{ \Phi_b(X(t))\Phi_b(X(t'))\rho_b(t_0) \right\},\qquad (2.49)$$

yields

$$\frac{\partial}{\partial t}\rho_s^{(I)}(t) = -\int_{t_0}^{t} dt' K_s(Q(t))K_s(Q(t'))\rho_s^{(I)}(t')C(t - t')$$

$$+ \int_{t_0}^{t} dt' K_s(Q(t))\rho_s^{(I)}(t')K_s(Q(t'))C^*(t - t')$$

$$+ \int_{t_0}^{t} dt' K_s(Q(t'))\rho_s^{(I)}(t')K_s(Q(t))C(t - t')$$

$$- \int_{t_0}^{t} dt' \rho_s^{(I)}(t')K_s(Q(t'))K_s(Q(t)))C^*(t - t').\qquad (2.50)$$

Contracting the terms above leads to

$$\frac{\partial}{\partial t}\rho_s^{(I)}(t) = -\int_{t_0}^{t} dt' \left[K_s(Q(t)), C(t - t')K_s(Q(t'))\rho_s^{(I)}(t') \right]$$

$$+ \int_{t_0}^{t} dt' \left[K_s(Q(t)), C^*(t - t')\rho_s^{(I)}(t')K_s(Q(t')) \right].\qquad (2.51)$$

The integral on the right-hand side of Eq. (2.51) cannot be evaluated analytically due to the unknown behaviour of $\rho_s^{(I)}(t')$. However, due to the system-bath interaction the system will loose the memory of its evolution in the bath after a characteristic bath correlation time τ_b. Thus one can approximate $\rho_s^{(I)}(t')$ by $\rho_s^{(I)}(t)$, assuming that τ_b is much shorter than the characteristic time scale of the system dynamics. This approximation is known as the Markov approximation [31]. Furthermore, the bath correlation function $C(t-t')$ can be approximated by 0 for $t-t' > \tau_b$ in the framework of the Markov approximation as all correlations are dephasing within τ_b. Therefore, the upper bound of the integral over t' can be extended to infinity. Applying the Markov approximation to Eq. (2.51) leads to

$$
\frac{\partial}{\partial t}\rho_s^{(I)}(t) = - \left[K_s(Q(t)), \int_{t_0}^{\infty} dt' C(t-t') K_s(Q(t')) \rho_s^{(I)}(t) \right]
$$
$$
+ \left[K_s(Q(t)), \rho_s^{(I)}(t) \int_{t_0}^{\infty} dt' C^*(t-t') K_s(Q(t')) \right]. \quad (2.52)
$$

Since the bath correlation function $C(t-t')$ is stationary and depends only on the difference of $\tau = t - t'$, one can rewrite the correlation function as

$$
C(\tau) = \mathrm{tr_{bath}} \left\{ \Phi_b(X(\tau)) \Phi_b(X(t_0)) \rho_b(t_0) \right\}. \quad (2.53)
$$

Introducing the operator

$$
\Lambda = \int_{t_0}^{\infty} d\tau C(\tau) K_s(Q), \quad (2.54)
$$

the corresponding QME in the Heisenberg picture is given by

$$
\frac{\partial}{\partial t}\rho_s(t) = - i \left[H_s, \rho_s(t) \right] - \left[K_s(Q), \Lambda \rho_s(t) - \rho_s(t) \Lambda^\dagger \right]. \quad (2.55)
$$

Note again that the approximations which are necessary to derive Eq. (2.55) restrict its applicability to systems with a weak system-bath interaction and a bath correlation time which is short in comparison to the time scale of the system dynamics.

2.2.2. Hierarchy equations of motion

In contrast to the QME approach discussed in the previous section, the HEOM formalism [34, 50–54] provides an in principle exact set of equations of motion for the reduced density matrix in the presence of arbitrary system-bath interactions. The basic idea of the HEOM approach is to mimic the system-bath interaction via an infinite hierarchy of auxiliary density matrices (ADM). Due to its computational demands the HEOM method is only applicable to rather small systems, composed, e.g., of a few chromophores, in a complex environment. Nevertheless, it can serve as a benchmark method to test the validity of more approximate methods.

The HEOM can be derived via path integral calculus utilising the Feynman-Vernon influence functional formalism [34, 50–52]. Note that Shao and co-workers [53, 54] proposed an alternative method to derive the HEOM. This approach separates the system and bath degrees of freedom via stochastic fields and leads to the same equations of motion as the influence functional formalism.

In analogy to Eq. (2.38), assuming again an initial factorization of $\rho(t_0)$ into system and bath parts, cf. Eq. (2.44), the formal time evolution of the reduced density matrix $\rho_s(t)$ is given by

$$\rho_s(t) = \mathrm{tr_{bath}} \left\{ U(t,t_0)\rho(t_0)U^\dagger(t,t_0) \right\} = \tilde{\mathcal{U}}(t,t_0)\rho_s(t_0). \qquad (2.56)$$

The time propagator in Liouville space $\tilde{\mathcal{U}}(t,t_0)$ still depends on the system as well as on the system-bath interaction part of the full system Hamiltonian. However, a complete separation of the system and bath DOFs can be achieved by virtue of the Feynman-Vernon influence functional formalism [55] rewriting $\tilde{\mathcal{U}}(t,t_0)$ in path integral representation. The time evolution operator $U(t,t_0)$ in path integral representation is defined as [31]

$$U(t,t_0) = \int_{\alpha_0}^{\alpha_t} \mathcal{D}\alpha \, e^{iS[\alpha]}. \qquad (2.57)$$

Here, α denotes arbitrary paths in phase space with fixed starting and end points α_0 and α_t, respectively. The action functional $S[\alpha]$ describes the evolution of the full system. According to Eq. (2.1) it can be decomposed into a system, a bath, and a system-bath interaction part. Thus, the time evolution of the full density matrix in path integral representation is given by [52]

$$
\rho(\alpha_t, \alpha_t', t) = \int_{\alpha_0}^{\alpha_t} \mathcal{D}\alpha \int_{\alpha_0'}^{\alpha_t'} \mathcal{D}\alpha' \, \mathrm{e}^{\mathrm{i}(S_\mathrm{s}[\alpha] + S_\mathrm{b}[\alpha] + S_\mathrm{s-b}[\alpha])} \rho(\alpha_0, \alpha_0', t_0)
$$

$$
\times \, \mathrm{e}^{-\mathrm{i}(S_\mathrm{s}[\alpha'] + S_\mathrm{b}[\alpha'] + S_\mathrm{s-b}[\alpha'])}
$$

$$
= \mathcal{U}(\alpha_t, \alpha_t', t; \alpha_0, \alpha_0', t_0)\rho(\alpha_0, \alpha_0', t_0). \tag{2.58}
$$

Taking the trace over the bath DOFs leads to [51]

$$
\tilde{\mathcal{U}}(\alpha_t, \alpha_t', t; \alpha_0, \alpha_0', t_0) = \int_{\alpha_0}^{\alpha_t} \mathcal{D}\alpha \int_{\alpha_0'}^{\alpha_t'} \mathcal{D}\alpha' \, \mathrm{e}^{\mathrm{i}S_\mathrm{s}[\alpha]} \mathcal{F}[\alpha, \alpha'] \mathrm{e}^{-\mathrm{i}S_\mathrm{s}[\alpha']}, \tag{2.59}
$$

where the system-bath interaction is fully covered by the Feynman-Vernon influence functional $\mathcal{F}[\alpha, \alpha']$ [55]. Note that while working in Liouville space it is necessary to keep track of the order of the terms as for instance the time evolution operator does not commute with the density operator.

Using the Caldeira-Leggett model for the system-bath interaction, Eq. (2.32), and restricting the description to a single system operator Q, the influence functional can be expressed as [52]

$$
\mathcal{F}[\alpha, \alpha'] = \exp\left(-\int_{t_0}^{t} d\tau \, \mathfrak{Q}\mathfrak{L} \right), \tag{2.60}
$$

with

$$
\mathfrak{Q} = Q[\alpha(\tau)] - Q[\alpha'(\tau)] \tag{2.61}
$$

and

$$\mathfrak{L} = \tilde{\Lambda}[\alpha(\tau)] - \tilde{\Lambda}^{\dagger}[\alpha'(\tau)]. \tag{2.62}$$

Here, the operator

$$\tilde{\Lambda}[\alpha(t)] = \int\limits_{t_0}^{t} \mathrm{d}\tau\, C(t-\tau)Q[\alpha(\tau)] \tag{2.63}$$

characterises the interaction of system and bath DOFs in terms of the bath correlation function $C(t-\tau)$, Eq. (2.49), including memory effects. Note that $\tilde{\Lambda}$ is actually the non-Markovian equivalent of the operator Λ defined by Eq. (2.54). Usually, $C(t-\tau)$ has a rather complex form depending on the actual system. However, an exponential form of the correlation function, i.e.

$$C(t) = \sum_{k=1}^{\mathcal{K}} C_k(t_0)\,\mathrm{e}^{-\gamma_k t}, \tag{2.64}$$

is required for the construction of the HEOM to avoid non-hierarchical terms [52]. In practice, $C(t)$ needs to be parametrized to fulfil this constraint. For simplicity a mono-exponentially decaying bath correlation function, i.e.

$$C(t) = C(t_0)\,\mathrm{e}^{-\gamma t}, \tag{2.65}$$

will be assumed in the following.

Taking the derivative of $\rho_{\mathrm{s}}(t)$ with respect to time leads formally to an equation of motion, which describes the time evolution of the system DOFs and incorporates the influence of the bath. Therefore, one needs to evaluate the derivative of the time evolution operator $\tilde{\mathcal{U}}(\alpha_t, \alpha'_t, t; \alpha_0, \alpha'_0, t_0)$, which in turn contains the derivatives of the action term and the influence functional $\mathcal{F}$. Whereas the former, describing the evolution of the unperturbed system, is given by

$$\frac{\partial}{\partial t}\,\mathrm{e}^{\mathrm{i}S_{\mathrm{s}}[\alpha]} = -\,\mathrm{i}H_{\mathrm{s}}\,\mathrm{e}^{\mathrm{i}S_{\mathrm{s}}[\alpha]}, \tag{2.66}$$

the latter, containing the effects of the system-bath interaction, reads, cf. Eq. (2.60),

$$\frac{\partial}{\partial t}\mathcal{F} = -\mathfrak{Q}\mathfrak{L}\mathcal{F}. \tag{2.67}$$

Equation (2.67) provides no closed solution as there exists no analytical expression for $\mathfrak{L}$ due to the memory effects contained in $\tilde{\Lambda}$, cf. the discussion of Eq. (2.51) and Ref. [52]. Applying the Markov approximation to Eq. (2.67) would reproduce the QME, Eq. (2.55), immediately. However, introducing the so-called auxiliary influence functional $\mathcal{F}_1$ one obtains the formal non-Markovian solution

$$\frac{\partial}{\partial t}\mathcal{F} = -\,\mathrm{i}\,[\mathfrak{Q}(-\,\mathrm{i}\mathfrak{L})]\,\mathcal{F} = -i\mathfrak{Q}\mathcal{F}_1, \tag{2.68}$$

which still provides no closed solution due to the aforementioned reasons. Nevertheless, a closed and formally exact solution of the problem can be obtained by taking the derivatives of $\mathcal{F}$ up to nth order. This leads to an infinite hierarchy of higher auxiliary influence functionals $\mathcal{F}_n$. Considering the time dependence of $\mathfrak{L}$, the first derivative of $\mathcal{F}_1$, i.e. the second derivative of $\mathcal{F}$, is given by

$$\begin{aligned}\frac{\partial}{\partial t}\mathcal{F}_1 &= -\,\mathrm{i}\left(\frac{\partial}{\partial t}\mathfrak{L}\right)\mathcal{F} - \mathrm{i}\mathfrak{Q}\mathcal{F}_2 \\ &= -\,\mathrm{i}\{C(t_0)Q - C^*(t_0)Q\}\mathcal{F} - \gamma\mathcal{F}_1 - \mathrm{i}\mathfrak{Q}\mathcal{F}_2,\end{aligned} \tag{2.69}$$

and the derivative of $\mathcal{F}_n$, i.e. the $(n+1)$th derivative of $\mathcal{F}$, by

$$\frac{\partial}{\partial t}\mathcal{F}_n = -\,\mathrm{i}\{C(t_0)Q - C^*(t_0)Q\}\mathcal{F}_{n-1} - n\gamma\mathcal{F}_n - \mathrm{i}\mathfrak{Q}\mathcal{F}_{n+1}. \tag{2.70}$$

Note that the time derivatives of $\mathfrak{L}$ in general could give rise to non-hierarchical terms. Due to the special choice of the bath correlation function, Eq. (2.65), i.e. the exponential shape of the function, these terms obey a hierarchical form as well. The leading order of the system-bath coupling of the auxiliary influence functional $\mathcal{F}_n$ is $2n$ [52]. Thus, in addition to memory effects also higher orders of the system-bath coupling are included in the HEOM formalism.

The obtained infinite hierarchy of influence functionals leads to an infinite hierarchy of auxiliary reduced density matrices,

$$\rho_n(t) = \tilde{\mathcal{U}}_n \rho_n(t_0), \tag{2.71}$$

where the auxiliary time evolution operator $\tilde{\mathcal{U}}_n$ is obtained by replacing the influence functional $\mathcal{F}$ in Eq. (2.59) by the corresponding auxiliary influence functional $\mathcal{F}_n$. The time evolution of the reduced density matrix and the ADM can be expressed in terms of a system of coupled differential equations, which is given by

$$\frac{\partial}{\partial t}\rho_n = -\left(\mathrm{i}\mathcal{L} + n\gamma\right)\rho_n - \mathrm{i}\left[Q, \rho_{n+1}\right]$$
$$- \mathrm{i}n\left[C(0)Q\rho_{n-1} - C^*(0)\rho_{n-1}Q\right]. \tag{2.72}$$

Here, $\mathcal{L}\bullet = [H_\mathrm{s}, \bullet]$ denotes the Liouville superoperator. The initial conditions for the reduced density matrices are $\rho_0(t_0) = \rho_\mathrm{s}(t_0)$ and $\rho_{n>0}(t_0) = 0$. Note that the density matrix ρ_0 represents the reduced system density matrix ρ_s. The ADM represent the influence of the bath and thus the memory effects, but contain no direct information on the system.

So far $C(t)$ was assumed to be a mono-exponential function, which is rather unlikely for real systems. However, the only constraint for the bath-correlation function, which is necessary to avoid non-hierarchical terms, is the aforementioned general exponential form of $C(t)$. Following the presented construction algorithm, using the general expansion of the correlation function, Eq. (2.64), leads to the generalized HEOM [52]

$$\frac{\partial}{\partial t}\rho_\mathbf{n} = -\mathrm{i}\mathcal{L}\rho_\mathbf{n} - \gamma_\mathbf{n}\rho_\mathbf{n} + \rho_\mathbf{n}^{(+)} + \rho_\mathbf{n}^{(-)}. \tag{2.73}$$

The index array $\mathbf{n} = (n_1, ..., n_\mathcal{K})$ characterizes the leading order of system-bath coupling $2n_k$ of each component k of the correlation function. The contributions $\rho_\mathbf{n}^{(+)}$ and $\rho_\mathbf{n}^{(-)}$ are given by

$$\rho_\mathbf{n}^{(+)} = -\mathrm{i}\sum_{k=1}^{\mathcal{K}}\left[Q_k, \rho_{\mathbf{n}_k^+}\right] \tag{2.74}$$

and

$$\rho_{\mathbf{n}}^{(-)} = -\,\mathrm{i} \sum_{k=1}^{\mathcal{K}} n_k \left(c_k Q_k \rho_{\mathbf{n}_k^-} - c_k^* \rho_{\mathbf{n}_k^-} Q_k \right). \qquad (2.75)$$

Here, the index array $\mathbf{n}_k^{\pm} = (n_1, n_2, ..., n_k \pm 1, ..., n_{\mathcal{K}})$ denotes the change of the kth component of the index array $\mathbf{n}$. Note that the components $k = 1, ..., \mathcal{K}$ of the correlation function do not necessarily need to be associated with the same interaction. Instead, they might, e.g., account for the interaction of the same environment with different subsystems or for the interaction of different environments with the same subsystem. In the former case the system part of the system-bath Hamiltonian Q_ζ, cf. Eq. (2.32), differs for the different subsystems, whereas in the latter case the correlation functions for the different environments differ. For instance, let us assume a molecular dimer system, in which the system operators are defined as $Q_{\zeta=1} = |e_1\rangle\langle e_1|$ and $Q_{\zeta=2} = |e_2\rangle\langle e_2|$ and each monomer is interacting with the same environment. The correlation function shall be approximated by two exponential terms with the prefactors C_1 and C_2. Therefore, the overall correlation function consists of four terms ($\mathcal{K} = 4$). The corresponding HEOM coefficients c_k and operators Q_k are given by

$$\begin{aligned}
c_1 &= C_1 & Q_1 &= |e_1\rangle\langle e_1| \\
c_2 &= C_2 & Q_2 &= |e_1\rangle\langle e_1| \\
c_3 &= C_1 & Q_3 &= |e_2\rangle\langle e_2| \\
c_4 &= C_2 & Q_4 &= |e_2\rangle\langle e_2|.
\end{aligned}$$

According to the index array $\mathbf{n}$, the prefactor of the direct influence term is given by

$$\gamma_{\mathbf{n}} = \sum_{k=1}^{\mathcal{K}} n_k \gamma_k. \qquad (2.76)$$

The hierarchy size $\mathcal{S}$ is defined by two dimensions, namely the number of expansion terms of the correlation function and the number of layers, where the Nth hierarchy layer consists of all density matrices with the same order of the system-bath coupling, i.e.

$$N = \sum_k^{\mathcal{K}} n_k. \tag{2.77}$$

The first three layers of a hierarchy with $\mathcal{K} = 2$ are shown in Fig. 2.2. According to the equations of motion, Eq. (2.73), the propagation of the reduced density matrix $\rho_s(t) = \rho_{00}(t)$ requires the auxiliary density matrices $\rho_{10}(t)$ and $\rho_{01}(t)$, which form the first hierarchy layer. The propagation of the latter matrices requires beside $\rho_{00}(t)$ additionally the second layer matrices and so on. Thus, the formally exact solution would require an infinite number of ADM and the hierarchy needs to be truncated for practical purposes.

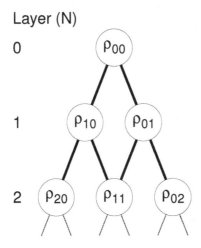

Figure 2.2.: Hierarchical scheme of the density matrices in the HEOM formalism for $\mathcal{K} = 2$.

The simplest approach to truncate the hierarchy is to limit the in principle infinite number of correlation function terms to a finite number $\mathcal{K}$ and the number of layers to a finite number $\mathcal{N}$. The overall number of reduced density matrices in the truncated hierarchy is then given by [35]

$$S(\mathcal{N}, \mathcal{K}) = \sum_{N=0}^{\mathcal{N}} \frac{(N + \mathcal{K} - 1)!}{N!(\mathcal{K} - 1)!}. \tag{2.78}$$

This truncation method is rather inefficient as usually large $\mathcal{K}$ and $\mathcal{N}$ are required to obtain converged results and the hierarchy size grows undesirably fast with increasing $\mathcal{K}$ and $\mathcal{N}$. More sophisticated truncation schemes have been proposed for both hierarchy dimensions [35, 36, 50, 56], which improve the convergence properties of the HEOM significantly. Applying the Markov approximation to the lowest hierarchy layer, Tanimura and co-workers developed a truncation method which takes into account the effects of the $(\mathcal{N} + 1)$th layer too [50]. A similar approach leads to a QME-like approximation of the residual part of the bath-correlation function [56]. Furthermore, introducing a normalized set of ADM, an efficient on-the-fly scaling algorithm has been developed [35], which reduces the numerical effort. This algorithm neglects all ADM whose largest element is below a certain threshold. An improved version of the HEOM, taking into account the residual approximation of the correlation function and the rescaling of the density matrices, is given by [36]

$$\frac{\partial}{\partial t} \rho_{\mathbf{n}} = -\mathrm{i}\mathcal{L}\rho_{\mathbf{n}} - \delta\mathcal{R}\rho_{\mathbf{n}} - \gamma_{\mathbf{n}}\rho_{\mathbf{n}} + \rho_{\mathbf{n}}^{(-)} + \rho_{\mathbf{n}}^{(+)}, \tag{2.79}$$

with

$$\rho_{\mathbf{n}}^{(+)} = -\mathrm{i}\sum_{k=1}^{\mathcal{K}} \sqrt{(n_k + 1)\sqrt{|c_k c_{k'}|}} \left[Q_k, \rho_{\mathbf{n}_k^+} \right], \tag{2.80}$$

$$\rho_{\mathbf{n}}^{(-)} = -\mathrm{i}\sum_{k=1}^{\mathcal{K}} \sqrt{\frac{n_k}{\sqrt{|c_k c_{k'}|}}} \left(c_k Q_k \rho_{\mathbf{n}_k^-} - c_{k'}^* \rho_{\mathbf{n}_k^-} Q_k \right) \tag{2.81}$$

and

$$\delta\mathcal{R}\bullet = \sum_{\zeta} \Delta_{\zeta} \left[Q_{\zeta}, \left[Q_{\zeta}, \bullet \right] \right]. \tag{2.82}$$

The index k' is defined such that $k' = l$ if $\gamma_l = \gamma_k^*$ and $Q_k = Q_l$ [36], which means that components belonging to a complex conjugated pair of bath correlation frequencies γ_k, cf. Eq. (2.64), are grouped. There exists always such a pair of conjugated frequencies for oscillatory components of the correlation function. For non-oscillatory components γ_k is real, thus $k' = k$. In contrast to the other summations, which are performed with respect to k, Eq. (2.80) and Eq. (2.81), the summation which accounts for the influence of the residual part of the correlation function, Eq. (2.82), is performed with respect to the physical system operator index ζ, cf. Eq. (2.32), to avoid a multiple counting of the residual contributions. The constant Δ_ζ can be calculated using the coefficients of the residual part of the correlation function, cf. App. A.

In addition to the conceptional developments, improved computational algorithms made it possible to study rather large systems like the B850 ring system of the light-harvesting antenna complex 2 of purple bacteria [37, 57, 58]. These algorithms are based on parallelized integration routines and the computation power of graphic processing units (GPUs).

2.3. Spectral density models

In the context of the QME and HEOM formalisms, which were discussed in the previous section, the key property, which determines the influence of the bath on the system dynamics, is the bath correlation function $C(t)$, Eq. (2.53). Unfortunately, in most cases $C(t)$ cannot be calculated directly as the evaluation of the right-hand side of Eq. (2.53) would require a quantum dynamics simulation of the full system. Such simulations are limited to rather small systems consisting of some hundred particles.

However, the Fourier transform of $C(t)$,

$$C(\omega) = \int_{-\infty}^{\infty} dt\, e^{i\omega t} C(t), \qquad (2.83)$$

contains all necessary information on the bath characteristics as well and can be determined approximately, e.g., by classical molecular dynamics simulations or experiments [31, 46]. The spectral distribution function $C(\omega)$ satisfies the detailed balance condition

$$C(-\omega) = e^{-\beta\omega}C(\omega). \tag{2.84}$$

Here, $\beta \equiv (k_B T)^{-1}$ denotes the inverse temperature. Introducing the symmetric and antisymmetric contributions of the spectral distribution function

$$C(\omega) = \frac{C^{(\pm)}(\omega)}{1 \pm e^{-\beta\omega}} \tag{2.85}$$

and the Bose-Einstein distribution function

$$n(\omega) = \frac{1}{e^{\beta\omega} - 1}, \tag{2.86}$$

the spectral distribution function can be expressed solely via its antisymmetric contribution, i.e.

$$C(\omega) = (1 + n(\omega))C^{(-)}(\omega). \tag{2.87}$$

Thus, the bath correlation function is given by

$$C(t) = \frac{1}{2\pi} \int_{-\infty}^{\infty} d\omega \, e^{-i\omega t}(1 + n(\omega))C^{(-)}(\omega). \tag{2.88}$$

Expanding the integral on the right-hand side of Eq. (2.88), using that $C^{(-)}$ is antisymmetric and

$$n(-\omega) = -e^{\beta\omega}n(\omega), \tag{2.89}$$

allows one to express $C(t)$ by the half-sided Fourier integral

$$C(t) = \frac{1}{2\pi} \int_{0}^{\infty} d\omega \left[e^{-i\omega t}(1 + n(\omega)) + e^{i\omega t}n(\omega) \right] C^{(-)}(\omega)$$

$$= \frac{1}{2\pi} \int_{0}^{\infty} d\omega \left[\cos(\omega t)\coth\left(\frac{\beta\omega}{2}\right) - i\sin(\omega t) \right] C^{(-)}(\omega). \tag{2.90}$$

However, the bath correlation functions $C(t)$ for the system-bath models introduced in Sec. 2.1 can be evaluated analytically. The bath correlation function for SBM I and IIa can be evaluated introducing creation and annihilation operators for the harmonic oscillator [31]. Due to the coupling of the different bath DOFs to the system DOFs in SBM IIb/c the calculation of the corresponding correlation functions requires more sophisticated methods, e.g., path integral techniques [46]. Note that the following discussion is therefore restricted to SBM I and IIa, whereas SBM IIb/c will be discussed separately later on.

The creation and annihilation operators, whose action on a complete set of harmonic oscillator eigenstates $|N_j\rangle$ is defined by

$$a_j^\dagger |N_j\rangle = \sqrt{N_j + 1} |N_j + 1\rangle \qquad (2.91)$$

and

$$a_j |N_j\rangle = \sqrt{N_j} |N_j - 1\rangle , \qquad (2.92)$$

are given by

$$a_j^\dagger(t) = \frac{1}{\sqrt{2}} \left(X_j(t) - iP_j(t) \right) \qquad (2.93)$$

and

$$a_j(t) = \frac{1}{\sqrt{2}} \left(X_j(t) + iP_j(t) \right) . \qquad (2.94)$$

Using the operators defined above, the position operator can be expressed as

$$X_j(t) = \frac{1}{\sqrt{2}} \left(a_j(t) + a_j^\dagger(t) \right) . \qquad (2.95)$$

Here, again X_j denotes a harmonic oscillator position operator and P_j the corresponding momentum operator. Inserting the bath part of the generic Caldeira-Leggett Hamiltonian, Eq. (2.32), which is given by

$$\Phi_b(X) = \sum_j c_{\zeta,j} X_j, \qquad (2.96)$$

in Eq. (2.49), and using Eq. (2.95), yields

$$C_\zeta(t) = \sum_j c_{\zeta,j}^2 \, \mathrm{tr_{bath}} \left\{ X_j(t) X_j(0) \rho_b(t_0) \right\}$$

$$= \sum_j \frac{1}{2} c_{\zeta,j}^2 \, \mathrm{tr_{bath}} \left\{ \left(a_j(t) + a_j^\dagger(t) \right) \left(a_j(0) + a_j^\dagger(0) \right) \rho_b(t_0) \right\}. \qquad (2.97)$$

Note that the description is restricted to fluctuations of the system coordinates, i.e. the overall correlation function separates into components associated with the system coordinates Q_ζ. For the present scenario the overall correlation function is given by

$$C(t) = \sum_\zeta C_\zeta(t). \qquad (2.98)$$

In general also terms of the type $C_{\zeta\zeta'}$ exist, which describe the fluctuations of the correlations between the system coordinates. However, these terms are neglected in the present work.

The trace on the right-hand site of Eq. (2.97) can be evaluated assuming that $\rho_b(t_0)$ describes an equilibrium state which corresponds to a thermal (Boltzmann) distribution of the bath harmonic oscillators, i.e. $\rho_b(t_0) = \mathcal{Z}^{-1} \exp(-\beta H_b)$, where $\mathcal{Z}$ denotes the partition function. Taking into account the time dependence of the creation and annihilation operators leads to

$$C_\zeta(t) = \sum_j \frac{1}{2} c_{\zeta,j}^2 \mathcal{Z}^{-1} \sum_N \langle N_j | \, \mathrm{e}^{-\mathrm{i}\omega_j t} a_j \left(a_j + a_j^\dagger \right) \mathrm{e}^{-\beta H_b} |N_j\rangle$$

$$+ \sum_j \frac{1}{2} c_{\zeta,j}^2 \mathcal{Z}^{-1} \sum_N \langle N_j | \, \mathrm{e}^{\mathrm{i}\omega_j t} a_j^\dagger \left(a_j + a_j^\dagger \right) \mathrm{e}^{-\beta H_b} |N_j\rangle$$

$$= \sum_j \frac{1}{2} c_{j,\zeta}^2 \sum_N f_{N_j} \left((1 + N_j) \, \mathrm{e}^{-\mathrm{i}\omega_j t} + N_j \, \mathrm{e}^{\mathrm{i}\omega_j t} \right) \qquad (2.99)$$

with $f_{N_j} = \mathcal{Z}^{-1}\exp(-\beta E_{N_j})$. The sum over $f_{N_j} N_j$ is just the definition of the Bose-Einstein distribution function [31], Eq. (2.86), and thus the bath correlation function is given by

$$C_\zeta(t) = \sum_j \frac{1}{2} c_{\zeta,j}^2 \left[(1 + n(\omega_j))\, e^{-i\omega_j t} + n(\omega_j)\, e^{i\omega_j t} \right] \qquad (2.100)$$

and the corresponding spectral distribution function by

$$C_\zeta(\omega) = \sum_j \pi c_{\zeta,j}^2 \left[(1 + n(\omega_j))\delta(\omega - \omega_j) + n(\omega_j)\delta(\omega + \omega_j) \right]. \qquad (2.101)$$

Introducing the spectral density

$$J_\zeta(\omega) = \frac{\pi}{2} \sum_j c_{\zeta,j}^2 \delta(\omega - \omega_j), \qquad (2.102)$$

using the properties of the δ-distribution and Eq. (2.89), the spectral distribution function Eq. (2.101) can be expressed via the spectral density as

$$C_\zeta(\omega) = 2[1 + n(\omega)][J_\zeta(\omega) - J_\zeta(-\omega)]. \qquad (2.103)$$

Note that in analogy to Eq. (2.98) the overall spectral density of the system reads

$$J(\omega) = \sum_\zeta J_\zeta(\omega). \qquad (2.104)$$

The corresponding antisymmetric contribution of $C_\zeta(\omega)$ is given by

$$C_\zeta^{(-)}(\omega) = 2[J_\zeta(\omega) - J_\zeta(-\omega)]. \qquad (2.105)$$

Inserting Eq. (2.105) into Eq. (2.90) leads to

$$C_\zeta(t) = \langle \Phi_{\mathrm{b}}(X(t))\Phi_{\mathrm{b}}(X(t_0))\rangle$$

$$= \frac{1}{\pi} \int\limits_0^\infty \mathrm{d}\omega \left[\cos(\omega t)\coth\left(\frac{\beta\omega}{2}\right) - i\sin(\omega t) \right] J_\zeta(\omega). \qquad (2.106)$$

This equation provides the necessary link between the spectral density function $J(\omega)$, which can be obtained by various theoretical and experimental techniques, with the bath correlation function $C(t)$, needed for quantum dynamics simulations. Note that $J(\omega)$ is defined only for positive values of ω, thus it is positive by definition. Approximations for $J(\omega)$ of natural systems are provided, e.g., by fluorescence line-narrowing and absorption experiments [9, 13] or a combination of molecular dynamics and electronic structure calculations [23, 59]. There exists also a number of model spectral densities, which are frequently used if no experimental or simulation data are available [60].

The bath-correlation function for a specific system-bath model can be evaluated by inserting the corresponding bath part of H_{s-b} instead of the generic Caldeira-Leggett one in Eq. (2.49) and following the presented algorithm. This leads for SBM Ib ($\zeta = m$, $j = i$, cf. Eq. (2.34)) to the monomeric spectral density

$$J_m^{(\text{Ib})}(\omega) = \frac{\pi}{2} \sum_i c_{i,m}^2 \delta(\omega - \omega_i) \tag{2.107}$$

and for SBM IIa, cf. Eq. (2.35), to

$$J_m^{(\text{IIa})}(\omega) = \pi \sum_\xi \omega_{\xi,m}^2 S_{\xi,m} \delta(\omega - \omega_{\xi,m})$$
$$+ \frac{\pi}{2} \sum_i c_{i,m}^2 \delta(\omega - \omega_i). \tag{2.108}$$

The aforementioned spectral densities are defined in terms of sums of delta functions weighted by some coefficients. However, any macroscopic system will in practice have a continuous spectral density due to the, at least for the environmental contribution, dense spectrum of DOFs. The environmental part of the spectral densities can be described by so-called Ohmic spectral densities, which are characterised by a linear rise of $J(\omega)$ for small frequencies and an exponential or Lorentzian cut-off. For example, the monomeric Ohmic spectral density with exponential cut-off is given by

$$J_m(\omega) = \Theta(\omega)\eta_m \omega \, e^{-\frac{\omega}{\gamma_m}}, \tag{2.109}$$

where the Heaviside function Θ assures that $J_m(\omega) = 0$ for $\omega < 0$. η_m represents the strength of the system-bath interaction and the cut-off term $\exp(-\omega/\tilde{\gamma}_m)$ determines the shape of $J_m(\omega)$ for values of ω which are similar to or larger than the cut-off frequency $\tilde{\gamma}_m$. The monomeric Ohmic spectral densitiy with a Lorentzian cut-off,

$$J_m(\omega) = \Theta(\omega)\eta_m\omega\frac{\tilde{\gamma}_m^2}{\omega^2 + \tilde{\gamma}_m^2}, \tag{2.110}$$

is usually called Debye spectral densitiy. The monomeric correlation function associated with the Debye spectral density can be calculated analytically using Eq. (2.106) and the residue theorem. One obtains [46]

$$\begin{aligned}
C_m(t) = &\frac{\eta_m\tilde{\gamma}_m^2}{2}\left(\cot\left(\frac{\beta\tilde{\gamma}_m}{2}\right) - i\right)e^{-\tilde{\gamma}_m t} \\
&+ \frac{2\eta_m\tilde{\gamma}_m^2}{\beta}\sum_{\nu=1}^{\infty}\frac{\gamma_\nu\,e^{-\gamma_\nu t}}{\gamma_\nu^2 - \tilde{\gamma}_m^2}.
\end{aligned} \tag{2.111}$$

Here, the summation over the Matsubara frequencies, $\gamma_\nu = 2\pi\beta^{-1}\nu$, stems from the poles of the Bose-Einstein distribution function, i.e. the hyperbolic cotangent in Eq. (2.106), whereas the first term stems from the pole of the spectral density itself [46]. Note that the Ohmic and Debye spectral densities describe the environmental part of the overall spectral density fairly well, but provide a rather crude approximation for the intra-molecular part as the latter is usually not continuous. Comparing the intra-molecular monomeric spectral density,

$$J_m^{(\text{int.})}(\omega) = \pi\sum_\xi \omega_{\xi,m}^2 S_{\xi,m}\delta(\omega - \omega_{\xi,m}), \tag{2.112}$$

and the shifted oscillator Hamiltonian, Eq. (2.19), one notes further that the reorganization energy, associated with the relaxation of the intra-molecular modes after a vertical transition from the monomeric ground to the excited state, can be evaluated by the integral

$$\frac{1}{\pi} \int_0^\infty d\omega \frac{J_m^{(\text{int.})}(\omega)}{\omega} = \frac{1}{\pi} \int_0^\infty d\omega \sum_\xi \frac{\omega_{\xi,m}^2 S_{\xi,m}}{\omega_{\xi,m}} \delta(\omega - \omega_{\xi,m})$$

$$= \sum_\xi S_{\xi,m} \omega_{\xi,m}. \qquad (2.113)$$

Accordingly, the corresponding sum of the monomeric Huang-Rhys factors is given by

$$\frac{1}{\pi} \int_0^\infty d\omega \frac{J_{\xi,m}(\omega)}{\omega^2} = \sum_\xi S_{\xi,m}. \qquad (2.114)$$

These properties of the spectral densities can be generalized, i.e. the overall reorganization energy is given by

$$E_{\text{reorg}} \equiv \frac{1}{\pi} \int_0^\infty d\omega \frac{J(\omega)}{\omega}, \qquad (2.115)$$

and the overall coupling strength of the system to the bath by

$$\tilde{S} \equiv \frac{1}{\pi} \int_0^\infty d\omega \frac{J(\omega)}{\omega^2}. \qquad (2.116)$$

So far a possible coupling between the different bath DOFs like in SBM IIb/c was neglected. The evaluation of the corresponding correlation functions requires, as mentioned above, the use of, e.g., path integral techniques. For instance, the monomeric spectral density for SBM IIb, which corresponds to the frequently used Multi-mode Brownian oscillator (MBO) model, is given by [46]

$$J_m^{(\text{IIb})}(\omega) = \sum_\xi S_{\xi,m} \omega_{\xi,m}^2 \frac{\omega \omega_{\xi,m} \gamma_{\xi,m}(\omega)}{(\omega^2 - \omega_{\xi,m}^2)^2 + \omega^2 \gamma_{\xi,m}^2(\omega)}. \qquad (2.117)$$

Here, the function $\gamma(\omega)$, which accounts for the interaction between the bath DOFs, is given by

$$\gamma_{\xi,m}(\omega) = \frac{\pi}{2} \sum_i c_{i,\xi m}^2 \, \delta(\omega - \omega_i). \tag{2.118}$$

The HEOM formalism, Sec. 2.2.2, requires a parametrization of the correlation function into a series of exponentials. Note that this parametrization requires a constant $\gamma_{\xi,m}(\omega)$, i.e. $\gamma_{\xi,m}(\omega)$ needs to be approximated by a characteristic frequency $\tilde{\gamma}_{\xi,m}$. This approximation is rather severe as it leads to the non-physical picture that there exists a practically dense spectrum of bath modes, which couple equally strong to arbitrary intra-molecular modes.

Assuming a constant $\gamma_{\xi,m}(\omega)$, inserting Eq. (2.117) into Eq. (2.106) and applying the residue theorem, leads to

$$C_m^{(\mathrm{IIb})}(t) = \sum_\xi \Xi_{\xi,m} \left(- e^{-\Omega_{\xi,m}^{(+)} t} \Upsilon_{\xi,m}^{(+)} + e^{-\Omega_{\xi,m}^{(-)} t} \Upsilon_{\xi,m}^{(-)} \right)$$
$$- \sum_\xi \Psi_{\xi,m} \sum_{\nu=1}^\infty \Gamma_{\xi,m}^{(\nu)}. \tag{2.119}$$

The parameters $\Xi_{\xi,m}$, $\Upsilon_{\xi,m}^{(\pm)}$, $\Psi_{\xi,m}$ and $\Gamma_{\xi,m}^{(\nu)}$ are defined as

$$\Xi_{\xi,m} = \frac{S_{\xi,m} \omega_{\xi,m}^3}{2 \Theta_{\xi,m}}, \tag{2.120}$$

$$\Upsilon_{\xi,m}^{(\pm)} = \coth\left(\frac{i\beta \Omega_{\xi,m}^{(\pm)}}{2} \right) - 1, \tag{2.121}$$

$$\Psi_{\xi,m} = \frac{4 S_{\xi,m} \omega_{\xi,m}^3 \tilde{\gamma}_{\xi,m}}{\beta}, \tag{2.122}$$

and

$$\Gamma_{\xi,m}^{(\nu)} = \frac{\gamma_n \, e^{-\gamma_n t}}{(\gamma_n^2 - \omega_{\xi,m}^2)^2 + \gamma_n^2 \tilde{\gamma}_{\xi,m}^2} \tag{2.123}$$

with

$$\Omega_{\xi,m}^{(\pm)} = \frac{\tilde{\gamma}_{\xi,m}}{2} \pm i\Theta_{\xi,m} \qquad (2.124)$$

and

$$\Theta_{\xi,m} = \sqrt{\omega_{\xi,m}^2 - \frac{\tilde{\gamma}_{\xi,m}^2}{4}}. \qquad (2.125)$$

The first two terms of Eq. (2.119) arise directly from the two poles of the spectral density, Eq. (2.117), whereas the third term arises from the poles of the Bose-Einstein distribution. These poles are as in Eq. (2.111) treated by the Matsubara scheme. Note that there exist other sum-over-poles schemes, like the Padé scheme, which have better convergence properties with respect to the number of terms required to model the correlation function with reasonable accuracy than the Matsubara scheme [36].

The correlation functions of the Debye and the MBO model have by construction the exponential form, Eq. (2.64), required by the HEOM formalism, cf. section 2.2.2. This is not the case for arbitrary spectral densities. Meier and Tannor [61] developed a numerical parametrization scheme which is based on the assumption that the actual dynamics of the system only depends on the value of the spectral density at the transition energies of the system and not on the shape of its individual components. They successfully parametrized an Ohmic spectral density, Eq. (2.109), by an expansion into three Lorentzian terms, using the parametrized spectral density,

$$\bar{J}(\omega) = \sum_{k=1}^{K} x_k \frac{\omega}{[(\omega + y_k)^2 + z_k^2][(\omega - y_k)^2 + z_k^2]}. \qquad (2.126)$$

Here, the set of parameters $\{x_k\}$, $\{y_k\}$ and $\{z_k\}$ for any given spectral density $J(\omega)$ can be obtained via minimizing the functional

$$\mathcal{F}[\{x_k\}, \{y_k\}, \{z_k\}, K] = \int_0^\infty d\omega |\bar{J}(\omega) - J(\omega)|. \qquad (2.127)$$

The correlation function according to the parametrized spectral density $\bar{J}(\omega)$ is given by

$$
\begin{aligned}
C(t) = \sum_{k=1}^{K} \frac{x_k}{y_k z_k} & [\coth(\frac{\beta}{2}(y_k + iz_k)) - 1] \, e^{(iy_k - z_k)t} \\
+ \sum_{k=1}^{K} \frac{x_k}{y_k z_k} & [\coth(\frac{\beta}{2}(y_k - iz_k)) - 1] \, e^{(-iy_k - z_k)t} \\
+ \frac{2\,i}{\beta} \sum_{\nu=1}^{\infty} & J(i\gamma_\nu) \, e^{-\gamma_\nu t}.
\end{aligned}
\tag{2.128}
$$

2.4. Calculation of optical spectra - response function formalism

The interaction between electromagnetic radiation and molecular systems provides valuable insights into their stationary and dynamical properties. There exists a variety of linear and nonlinear spectroscopy techniques, which are suitable to study, e.g., the energy transport in molecular aggregates. The key property, which connects the macroscopic response of the system to the applied electric field $\mathfrak{E}(t)$ with the underlying microscopic dynamics, is the polarization $P[\mathfrak{E}(t), t]$. In dipole approximation, i.e. the interaction of the system and the field is described by Eq. (2.24), $P[\mathfrak{E}(t), t]$ of a homogeneous sample is given by [31]

$$
P[\mathfrak{E}(t), t] = n_{\mathrm{mol}} \operatorname{tr}\{\rho(t)d\}.
\tag{2.129}
$$

Here, d denotes the microscopic dipole operator, Eq. (2.26), and n_{mol} stands for the volume density of the molecules in the sample volume. Note that $\mathfrak{E}(t)$, d and $P[\mathfrak{E}(t), t]$ are in general vectors, but for simplicity the vector character will be neglected in the following.

Whereas the microscopic dynamics of the system is described, e.g., by the Schrödinger equation or the equations of motion introduced in Sec. 2.2, the macroscopic connection between the field and the polarization is determined by the Maxwell equations, i.e.

$$\left(\frac{\partial^2}{\partial t^2} - c^2 \Delta\right) \mathfrak{E}(t) = -4\pi \frac{\partial^2}{\partial t^2} P[\mathfrak{E}(t), t]. \tag{2.130}$$

Here, c denotes the speed of light and Δ the Laplace operator. In principle, one needs to solve the coupled equations, given by the field equation and the system's equations of motion, self-consistently to describe the light-matter interaction exactly. However, often such a rigorous treatment is not necessary in the limit of weak system-field interaction [46], i.e. it is sufficient to solve first the equations of motion of the system driven by the external field and to calculate the corresponding macroscopic signal afterwards.

The functional $P[\mathfrak{E}(t), t]$ is in general a nonlinear functional of the electric field, but can be decomposed into its linear and nonlinear components. In the following sections the connection of the components of the polarization to linear and nonlinear spectroscopy techniques will be discussed.

2.4.1. Linear Response

Let us assume a linear relationship between the polarization and the electric field, i.e.

$$P = \chi \mathfrak{E}, \tag{2.131}$$

and that the electric field is given by

$$\mathfrak{E}(t) = \tilde{\mathfrak{E}}(t)\, e^{i\vec{k}\vec{r} - i\omega t} + \text{c.c.}. \tag{2.132}$$

Here, $\tilde{\mathfrak{E}}(t)$ represents the envelope function of the field with wave-vector $\vec{k}$. The absorption coefficient $\alpha(\omega)$ describing the decay of the field intensity inside the sample according to Beer's law is given by [31]

$$\alpha(\omega) = \frac{4\pi\omega}{nc} \mathfrak{Im}(\chi(\omega)), \tag{2.133}$$

where n denotes the index of refraction of the sample. The dielectric susceptibility $\chi(\omega)$ depends on the system properties and can be determined microscopically via the so-called linear response function $R^{(1)}(t)$. Assuming a weak field, which is usually the case in the experiment, it is convenient to expand Eq. (2.129) in powers of the electric field strength. This leads to [31]

$$P(t) = i \underbrace{\int_0^{\infty} d\tau\, \Theta(\tau) n_{\mathrm{mol}} \operatorname{tr} \left\{ \rho_{\mathrm{eq}} \left[d^{(I)}(\tau), d^{(I)}(0) \right] \right\}}_{R^{(1)}(t)} \mathfrak{E}(t - \tau). \quad (2.134)$$

Here, $\Theta(t)$ denotes again the Heaviside function and the dipole operator is given in the interaction representation, cf. Sec. 2.2.1, with respect to the system-field interaction. It is assumed further that the initial density matrix describes the equilibrium of the system, i.e. $\rho(t_0) = \rho_{\mathrm{eq}}$. Note that the polarization is given by the response function itself in the so-called impulsive limit, i.e. $\tilde{\mathfrak{E}}(t - \tau) = \tilde{\mathfrak{E}}\delta(t - \tau)$. Applying the convolution theorem, the Fourier transform of Eq. (2.134) is given by

$$P(\omega) = \chi(\omega)\mathfrak{E}(\omega) \quad (2.135)$$

with

$$\chi(\omega) = \int_{-\infty}^{\infty} dt\, e^{i\omega t} R^{(1)}(t). \quad (2.136)$$

Thus, the absorption coefficient is proportional to the Fourier transform of the linear response function, $R^{(1)}(t)$, i.e.

$$\alpha(\omega) \propto \int_0^{\infty} dt\, e^{i\omega t} \operatorname{tr} \left\{ \rho_{\mathrm{eq}} \left[d^{(I)}(\tau), d^{(I)}(0) \right] \right\}. \quad (2.137)$$

Expanding the commutator on the right-hand side of Eq. (2.137) in the Schrödinger picture leads to

$$\alpha(\omega) \propto \int_0^\infty dt\, e^{i\omega t}\, \mathrm{tr}\left\{\rho_{eq}(U^\dagger(t,t_0)dU(t,t_0)d - dU^\dagger(t,t_0)dU(t,t_0))\right\}$$

$$= \int_0^\infty dt\, e^{i\omega t}\, \mathrm{tr}\left\{dU(t,t_0)d\rho_{eq}U^\dagger(t,t_0) - dU(t,t_0)\rho_{eq}dU^\dagger(t,t_0)\right\}$$

$$= \int_0^\infty dt\, e^{i\omega t}\, \mathrm{tr}\left\{d\sigma^{(+)}(t) - d\sigma^{(-)}(t)\right\}. \tag{2.138}$$

The time evolution operator $U(t,t_0)$ is defined by Eq. (2.39). Note that the second term under the trace of the right-hand side of Eq. (2.138) gives rise to anti-resonant contributions and can be neglected [31]. Therefore, the absorption coefficient is given by

$$\alpha(\omega) \propto \int_0^\infty dt\, e^{i\omega t}\, \mathrm{tr}\left\{d\sigma^{(+)}(t)\right\}. \tag{2.139}$$

So far a homogeneous sample was considered. In the experiment this condition is not fulfilled as the molecules are in general randomly oriented and individual molecules are influenced differently by their environment. While the latter effect is partly taken into account by the system-bath model outlined in the previous sections, the former effect needs to be incorporated by averaging the spectra over the different orientations of the molecules with respect to the incoming field [62].

2.4.2. Nonlinear Response

In the previous section only the linear contribution of the polarization was considered. Expanding the polarization in powers of the field strength, i.e.

$$P = P^{(1)} + P^{(2)} + \dots \tag{2.140}$$

and extending the perturbative treatment of the expectation value of the polarization, the nth order component of the polarization is given by [46]

$$P^{(n)} = \int_0^\infty dt_n \dots dt_1 R^{(n)}(t_n, \dots, t_1)$$

$$\times \mathfrak{E}(t - t_n) \dots \mathfrak{E}(t - t_n - \dots - t_1). \qquad (2.141)$$

The nth order response function is defined as

$$R^{(n)}(t_n, \dots, t_1) = i^n n_{\text{mol}} \Theta(t_1) \dots \Theta(t_n) \operatorname{tr}\left\{\rho_{\text{eq}} \mathfrak{K}^{(n)}\right\} \qquad (2.142)$$

with

$$\mathfrak{K}^{(n)} = \left[\left[\left[\dots\left[d^{(I)}(t_n + \dots + t_1), d^{(I)}(t_{n-1} + \dots + t_1)\right]\dots\right],\right.$$

$$\left.\times d^{(I)}(t_1)\right], d^{(I)}(0)\right]. \qquad (2.143)$$

In analogy to the linear response function $R^{(1)}$ the nth order one, $R^{(n)}$, contains all necessary information to evaluate the corresponding signals. The electric field, $\mathfrak{E}(t)$, is in general defined by Eq. (2.132). However, according to the multi-pulse schemes commonly used in the experiments, it is convenient to separate the overall field into individual components corresponding to the different pulses, i.e.

$$\mathfrak{E}(t) = \sum_{j=1}^n \left(\tilde{\mathfrak{E}}_j(t) e^{i\vec{k}_j \vec{r} - i\omega_j t} + \tilde{\mathfrak{E}}_j^*(t) e^{-i\vec{k}_j \vec{r} + i\omega_j t}\right). \qquad (2.144)$$

Note that the wave-vector and the carrier frequency of the signal field, resulting from the nonlinear polarization, need to obey the conditions

$$\vec{k}_s = \pm j_1 \vec{k}_1 \pm j_2 \vec{k}_2 \pm \dots \pm j_n \vec{k}_n \qquad (2.145)$$

and

$$\omega_s = \pm j_1 \omega_1 \pm j_2 \omega_2 \pm \dots \pm j_n \omega_n, \qquad (2.146)$$

due to the conservation of energy and momentum. The integer numbers j_i account for the possibility of multiple interactions of the sample with the same pulse. The phase-matching condition, Eq. (2.145), limits the observable signal in the experiment to certain contributions of the overall nonlinear signal. This gives rise to a variety of different nonlinear spectroscopy techniques focussing on special contributions.

Frequently, third order spectroscopy techniques are used to investigate the excitation energy transport in molecular aggregates and other processes. Such nonlinear spectroscopy methods are advantageous in comparison to linear spectroscopy techniques as they probe, due to the multiple interactions of the system with the field, directly the dynamics of the system. Beside the commonly used transient absorption scheme [63], the electronic 2D-spectroscopy technique [64, 65] provides a powerful tool to study the dynamics of rather complex systems [16, 66, 67]. In the former method two laser pulses, which are delayed with respect to each other, are applied to the sample, whereas in the latter a three-pulse scheme is used. The third-order response function, which describes both techniques, is given by

$$R^{(3)}(t_3, t_2, t_1) = i^3 n_{\text{mol}} \Theta(t_1)\Theta(t_2)\Theta(t_3)$$
$$\times \text{tr}\left\{\rho_{\text{eq}}\left[\left[\left[d^{(I)}(t_3 + t_2 + t_1), d^{(I)}(t_2 + t_1)\right], d^{(I)}(t_1)\right], d^{(I)}(0)\right]\right\}.$$
$$(2.147)$$

Expanding the commutators on the right-hand side of Eq. (2.147) leads to

$$R^{(3)}(t_3, t_2, t_1) = i^3 n_{\text{mol}} \Theta(t_1)\Theta(t_2)\Theta(t_3)$$

$$\times \sum_{i=1}^{4}(R_i(t_3, t_2, t_1) - R_i^*(t_3, t_2, t_1)). \qquad (2.148)$$

The individual components R_i, which correspond to different combinations of interactions, can be associated with physical processes like ground state bleaching, excited state absorption or stimulated emission and are given by

$$R_1(t_3, t_2, t_1) = \text{tr} \left\{ d^{(I)}(t_3 + t_2 + t_1) d^{(I)}(0) \rho_{\text{eq}} d^{(I)}(t_1) d^{(I)}(t_2 + t_1) \right\}$$
(2.149)

$$R_2(t_3, t_2, t_1) = \text{tr} \left\{ d^{(I)}(t_3 + t_2 + t_1) d^{(I)}(t_1) \rho_{\text{eq}} d^{(I)}(0) d^{(I)}(t_2 + t_1) \right\}$$
(2.150)

$$R_3(t_3, t_2, t_1) = \text{tr} \left\{ d^{(I)}(t_3 + t_2 + t_1) d^{(I)}(t_2 + t_1) \rho_{\text{eq}} d^{(I)}(0) d^{(I)}(t_1) \right\}$$
(2.151)

$$R_4(t_3, t_2, t_1) = \text{tr} \left\{ d^{(I)}(t_3 + t_2 + t_1) d^{(I)}(t_2 + t_1) d^{(I)}(t_1) d^{(I)}(0) \rho_{\text{eq}} \right\}.$$
(2.152)

Note that the terms under the trace are rearranged such that the last interaction is the leftmost one. This arrangement can be pictured with the commonly used double-sided Feynman diagram technique [46], which provides an intuitive picture of the evolution of the density matrix. In general the double-sided Feynman diagrams consist of two vertical lines representing the bra and the ket of the density matrix. Time is evolving from the bottom to the top of the diagram and interactions are represented by diagonal arrows, which have their origin at or are pointing to the solid bra or ket lines. Whereas arrows pointing to the bra or the ket lines indicate an excitation process, arrows which have their origin at the bra or the ket lines indicate a de-excitation process. Arrows pointing to the right represent an interaction with a field contribution of $\tilde{\mathfrak{E}}_j(t)\exp(i\vec{k}_j\vec{r} - i\omega_j t)$ and arrows pointing to the left an interaction with a field contribution of $\tilde{\mathfrak{E}}_j^*(t)\exp(-i\vec{k}_j\vec{r} + i\omega_j t)$. There exists a large set of Feynman diagrams describing all possible third-order interactions. However, using the phase matching condition, Eq. (2.145), assuming a fixed time ordering of the pulses and neglecting off-resonant terms, i.e. applying the rotating wave approximation [46], leads to a small subset of Feynman diagrams describing all processes, which might be observed using a particular spectroscopy technique for a given level scheme.

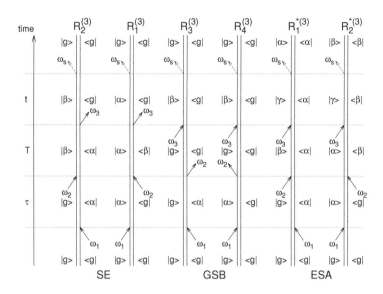

Figure 2.3.: Double-sided Feynman diagrams representing the interaction pathways leading to the rephasing and nonrephasing 2D-spectroscopy signals.

The aforementioned electronic 2D-spectroscopy technique, which applies three independent laser pulses to the sample, offers in principle eight possible phase-matching directions. However, only two directions are frequently used in the experiment, namely the photon echo direction $\vec{k}_s^{(\mathrm{rp})} = -\vec{k}_1 + \vec{k}_2 + \vec{k}_3$, which gives rise to the so-called rephasing Feynman diagrams, and the transient gradient direction $\vec{k}_s^{(\mathrm{nr})} = \vec{k}_1 - \vec{k}_2 + \vec{k}_3$, which yields the so-called non-rephasing diagrams.

All rephasing Feynman diagrams, Fig. 2.3, have in common that the first pulse creates a coherence of the type $|g\rangle\langle\alpha|$, whereas the third pulse creates a coherence of the type $|\beta\rangle\langle g|$ or $|\gamma\rangle\langle\alpha|$. Here, $|g\rangle$ denotes the ground state of the system, $|\alpha\rangle$ and $|\beta\rangle$ arbitrary one-exciton states and $|\gamma\rangle$ an arbitrary higher excited state. During the so-called coherence time τ between the first and the second pulse, the system accumulates a phase corresponding to the energy gap

between the states $|g\rangle$ and $|\alpha\rangle$. The second pulse drives the system
into a population of the excited state or another coherence. After the
so-called population time T between the second and the third pulse,
the system is driven into the final coherence by the third pulse. The
system accumulates again a phase during the so-called rephasing time
t, but now with a different sign counteracting the phase, accumulated
during the coherence time τ, and thus leading to a rephasing, which
depends on the system-bath interaction. In non-rephasing Feynman
diagrams the last coherence is of the same type as the first one and
therefore no rephasing occurs. The three rephasing and three non-
rephasing Feynman diagrams can also be classified according to the
process which they represent. Whereas the first two diagrams account
for stimulated emission (SE) and diagrams three and four for ground
state bleaching (GSB), the last two diagrams represent excited state
absorption (ESA) which is absent for the aggregate model introduced
in Sec. 2.1 as this model is restricted to singly excited states.

In the impulsive limit, i.e. $\tilde{\mathfrak{E}}_j(t) \propto \tilde{\mathfrak{E}}_j \delta(t)$, the rephasing and non-
rephasing 2D-spectra can be obtained via double Fourier transform
of the response function with respect to τ and t [68]. Note that the
rephasing and non-rephasing spectra include both absorptive and
dispersive features. The dispersive features can be eliminated by
adding both the non-rephasing and the rephasing spectra [69], thus
the absorptive 2D spectrum is given by [70, 71]

$$S(\omega_\tau, T, \omega_t) \propto \mathfrak{Im}\left(\int_0^\infty \mathrm{d}t \int_0^\infty \mathrm{d}\tau \left[e^{i\omega_\tau \tau + i\omega_t t} R_{\mathrm{nr}}^{(3)} + e^{-i\omega_\tau \tau + i\omega_t t} R_{\mathrm{rp}}^{(3)} \right]\right)$$

$$(2.153)$$

with

$$R_{\mathrm{rp}}^{(3)}(t_3, t_2, t_1) = R_2(t_3, t_2, t_1) + R_3(t_3, t_2, t_1) - R_1^*(t_3, t_2, t_1) \quad (2.154)$$

$$R_{\mathrm{nr}}^{(3)}(t_3, t_2, t_1) = R_1(t_3, t_2, t_1) + R_4(t_3, t_2, t_1) - R_2^*(t_3, t_2, t_1). \quad (2.155)$$

A detailed discussion of different features in 2D-spectra arising from
the underlying level structure is given, e.g., in Refs. [64, 65, 68]. Briefly,

the diagonal $\omega_\tau = \omega_t$ of the 2D-spectrum for $T = 0$ fs corresponds to the linear absorption spectrum, i.e. represents the level structure of the sample, whereas off-diagonal peaks indicate a coupling between different levels. Therefore, a change in the intensity distribution of the 2D-spectra for different population times T represents the dynamics of the system. For example a growth of the off-diagonal features below the diagonal indicates an energy transfer or relaxation between the levels corresponding to the respective diagonal features.

The rephasing and non-rephasing response functions can be calculated, analogously to the linear response function, Eq. (2.138), in the Schrödinger picture. Defining the propagator $\mathcal{G}(t_j, t_i)$ for the field-free propagation of the density matrix during the interval $[t_i, t_j]$, the response functions can be evaluated by

$$R_{\mathrm{rp}}^{(3)}(t_3, t_2, t_1) =$$
$$\mathrm{i}^3 \, \mathrm{tr} \left\{ d^{(-)} \mathcal{G}(t_3, t_2) \left[d^{(+)}, \mathcal{G}(t_2, t_1) \left[d^{(+)}, \mathcal{G}(t_1, t_0) \left[d^{(-)}, \rho_{\mathrm{eq}} \right] \right] \right] \right\}$$
(2.156)

$$R_{\mathrm{nr}}^{(3)}(t_3, t_2, t_1) =$$
$$\mathrm{i}^3 \, \mathrm{tr} \left\{ d^{(-)} \mathcal{G}(t_3, t_2) \left[d^{(+)}, \mathcal{G}(t_2, t_1) \left[d^{(-)}, \mathcal{G}(t_1, t_0) \left[d^{(+)}, \rho_{\mathrm{eq}} \right] \right] \right] \right\}.$$
(2.157)

Here, the operators $d^{(-)}$ and $d^{(+)}$ are defined as

$$d^{(-)} = \sum_m d_{g,e_m} |g\rangle\langle e_m| \qquad (2.158)$$
$$d^{(+)} = \sum_m d_{e_m,g} |e_m\rangle\langle g|. \qquad (2.159)$$

Note that the calculation of 2D-spectra is much more demanding from the computational point of view, as the propagations need to cover both the t_1 as well as the t_3 range.

3. Energy transfer in light-harvesting complexes

Energy migration within the photosynthetic machineries of algae, bacteria and plants has long been considered as a hopping process, which can be described via rate equations. Further, it was assumed that the structure of the photosynthetic units is optimized to maximize the quantum yield of the overall process. However, this simplified approach was questioned after the observation of long lasting coherent oscillations in the 2D-spectra of the FMO complex by Engel and co-workers in 2007 [16]. Since then the effect of coherence on the energy transport properties and its origin has been one of the major lines of investigation in the context of photosynthesis [18, 20, 21, 26–28, 72, 73]. The dynamics within light-harvesting antennae after an optical excitation is rather complex due to their sophisticated structure and the influence of the environment. Therefore, it is instructive to study, in addition to natural light-harvesting units, model systems to disentangle the dynamics which lead to the high quantum efficiency of these complexes.

In the following section the dynamics of a generic model dimer will be investigated using the HEOM method, cf. Sec. 2.2.2, implemented in the Rostock HEOM package, cf. App. A. Here, the issue of coherent oscillations as well as the general influence of the bath on the system dynamics in dependence on the system-bath interaction strength will be addressed. Next, the energy transfer in an octamer forming an energy funnel towards a sink will be studied as a model aggregate which resembles, e.g., the function of photosynthetic antennae, cf. Fig. 1.1. Finally, the energy transfer of the FMO complex itself will be modelled using a spectral density extracted from experimental data.

3.1. Dimer system

The dimer system represents the smallest possible molecular aggregate. Due to its low dimensionality it can be treated numerically rather exactly in comparison to larger aggregates. Hence the dimer is of particular interest as a model system which provides a reference for more complicated systems [20, 22, 24, 26, 27, 57, 71–78].

To investigate the dissipative exciton dynamics in dimer systems a generic model incorporating one intra-molecular mode modelled via a single mode MBO spectral density, Eq. (2.117), is used in the following. The electronic energy gap between the diabatic excited aggregate states ($\Delta E = E_{e_2} - E_{e_1} = 500$ cm^{-1}), cf. Eq. (2.4), is chosen such that it corresponds to the frequency of the intra-molecular mode $\omega_{1,1} = \omega_{1,2} = \omega_{\text{vib}} = 500$ cm^{-1}. This situation is rather typical for light-harvesting complexes, e.g., in the FMO complex there exists a mode with $\omega_{\text{vib}} = 180$ cm^{-1}, which is resonant to the gap energy between monomer three and four, cf. Sec. 3.3. This facilitates a strong mixing of the electronic and vibrational DOFs, which has been discussed as one possible origin of long-living oscillations in the 2D-spectra of FMO [21]. Note that all parameters except of the electronic state energies are assumed to be the same for both monomers, e.g., $S_{1,1} = S_{1,2} = S$, and therefore the monomer index m and the mode index ξ will be skipped. The temperature of the bath is chosen to be $T = 300$ K.

In order to study the influence of the vibronic coupling strength, i.e. the Huang-Rhys factor S, on the system dynamics, four different dimer scenarios are considered. They are characterised by means of their linear absorption spectra, population dynamics and 2D-spectra. The scenarios cover in particular the cases of weak (I and II) and strong (III and IV) vibronic coupling for two different Coulomb couplings, see Tab. 3.1, Each scenario will be investigated for two different inter-bath coupling strengths, namely $\tilde{\gamma} = 50$ cm^{-1} and $\tilde{\gamma} = 200$ cm^{-1}. Example inputs for the Rostock HEOM package are provided in app. B. Converged results for all scenarios were obtained with $\mathcal{K} = 2$ and $\mathcal{N} = 9$, cf. Sec. 2.2.2.

Table 3.1.: Huang-Rhys factors, Coulomb coupling strengths and ratios of the Coulomb coupling strength and vibrational frequency for the different dimer scenarios.

scenario	S	J [cm^{-1}]	J/ω_{vib}
I	0.05	250	0.5
II	0.05	750	1.5
III	0.5	250	0.5
IV	0.5	750	1.5

The linear absorption spectra for I and II, Fig. 3.1 panels a and b[1], show only a significant absorption at the energies corresponding to the electronic dimer, cf. Eq. (2.23), whereas the spectra for III and IV, panels c and d, show pronounced additional side peaks due to the strong vibronic coupling. In all cases the second adiabatic electronic state and the associated vibronic states carry the main part of the oscillator strength, cf. Fig. 3.4. This is characteristic for so-called H-aggregates and a direct consequence of the Coulomb coupling configuration ($J > 0$) [31]. Aggregates with $J < 0$, the so-called J-aggregates, show pronounced peaks at the low-energy site of the spectra, i.e. the first adiabatic state and the associated vibronic states carry the main part of the oscillator strength. In contrast to the spectra for I, II and IV, the spectrum for III, where the reorganization energy associated with the vibronic coupling is comparable to the Coulomb coupling strength, shows a rather complex shape. Increasing the inter-bath coupling strength $\tilde{\gamma}$ leads to a larger line width in all cases. As a consequence of the chosen system-bath model, cf. Sec. 2.1, this effect is again more pronounced for III and IV. The line width is determined by both, the coupling of the electronic DOFs to the primary bath, i.e. the Huang-Rhys factor S, and the coupling between the primary and secondary bath, i.e. $\tilde{\gamma}$. For I and II the small Huang-Rhys factor is limiting the interaction between the system

---

[1] Please visit www.springer.com and search for the author's name to access the chapters colored figures.

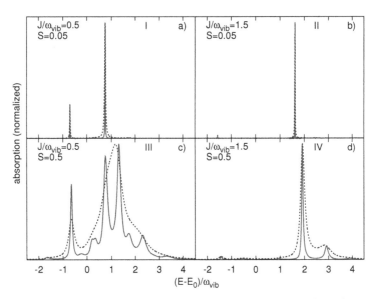

Figure 3.1.: Absorption spectra corresponding to the four dimer scenarios ($\tilde{\gamma} = 50$ cm^{-1} solid red line, $\tilde{\gamma} = 200$ cm^{-1} dashed blue line) calculated with the Rostock HEOM package using Eq. (2.139). The spectra are normalized such that the peak value of the dominant peak is equal to unity. The reference energy E_0 is equal to the mean of the diabatic electronic state energies ($E_0 = \Delta E/2$).

and the bath and therefore restricts the line width independently of the magnitude of the inter-bath coupling strength. In contrast for III and IV the interaction between electronic DOFs and primary bath is strong, i.e. the interaction between primary and secondary bath directly influences the decoherence rate of the system DOFs and therefore the line width of the spectra. Note that peaks with a significant vibrational contribution, cf. discussion of Fig. 3.4, i.e. the peaks corresponding to the vibronic progression in the panels c and d of Fig. 3.1, show a stronger line broadening for increasing $\tilde{\gamma}$ than peaks with mainly electronic character. This is due to the stronger interaction of vibrationally excited states with the bath in comparison to the vibrational ground state.

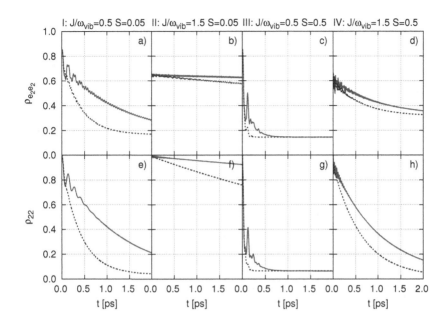

Figure 3.2.: Population dynamics of the highest diabatic (upper row) and adiabatic (lower row) states for the four dimer scenarios ($\tilde{\gamma} = 50\ \mathrm{cm}^{-1}$: solid red line, $\tilde{\gamma} = 200\ \mathrm{cm}^{-1}$: dashed blue line).

The population dynamics of the diabatic and adiabatic states, Fig. 3.2, after initial excitation of the highest adiabatic state is significantly influenced by the inter-bath coupling for all scenarios. Whereas for $\tilde{\gamma} = 50\ \mathrm{cm}^{-1}$ all populations except of those for II (panels b and f) feature pronounced oscillations over at least 500 fs, these oscillations are damped out rapidly for $\tilde{\gamma} = 200\ \mathrm{cm}^{-1}$. Additionally, the relaxation towards the first diabatic/adiabatic state is significantly faster for the stronger inter-bath coupling.

The population dynamics of the diabatic state $|e_2\rangle$, as well as of the adiabatic state $|2\rangle$, Eq. (2.22), for I, III and IV shows an exponential decay of the state population. This is in contrast to III, where the diabatic and adiabatic populations decay almost linearly. In general the population relaxation for the scenarios featuring a

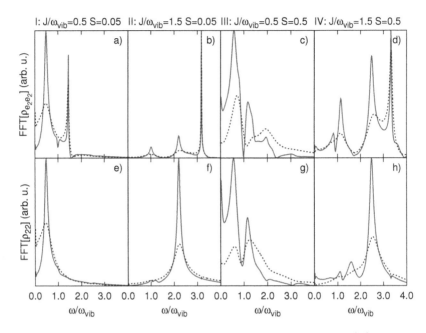

Figure 3.3.: Fourier spectra of the oscillatory component of the populations shown in Fig. 3.2 ($\tilde{\gamma} = 50$ cm^{-1}: solid red line, $\tilde{\gamma} = 200$ cm^{-1}: dashed blue line). The Fourier spectra are normalized such that the area under the curve is equal to unity.

strong vibronic coupling (III and IV) is faster than the one for the scenarios featuring a weak vibronic coupling (I and II). Comparing the population dynamics for I and II as well as those for III and IV, respectively, one further notices that the population relaxation is faster for the scenarios with a smaller Coulomb coupling. This is due to the fact that the energy gap between the adiabatic electronic states, cf. Eq. (2.23), for $J = 250$ cm^{-1} is closer to ω_{vib} than for $J = 750$ cm^{-1}. Thus, the value of the spectral density, which has its maximum at $\omega = \omega_{\text{vib}}$, at the transition energy between the states is larger in the former case than in the latter case. This leads to a stronger coupling to the bath for $J = 250$ cm^{-1}, which in return facilitates a faster relaxation for I and III in comparison to II and IV.

Comparing the diabatic and adiabatic population dynamics one notices that, although the general behaviour is the same for both, the adiabatic populations feature fewer oscillations than the corresponding diabatic ones. In order to characterise the oscillatory behaviour of the population dynamics in more detail, it is instructive to inspect the Fourier spectrum (FS) of the oscillations. It can be calculated by subtracting the decaying component, i.e.

$$P_{\text{osc}}(t) = P(t) - P_{\text{dec}}(t), \tag{3.1}$$

and taking the Fourier transform of the $P_{\text{osc}}(t)$. The decaying component $P_{\text{dec}}(t)$ of the population can be extracted, e.g., via an exponential fit, i.e.

$$P_{\text{dec}}(t) = P(0)\, e^{-\frac{t}{\tau}}. \tag{3.2}$$

According to the population dynamics shown in Fig. 3.2, the FS of the adiabatic populations, Fig. 3.3, show in general less features than the FS of the corresponding diabatic ones. In particular for I, II and IV the distinct peaks at the high frequencies in the FS of the diabatic populations are absent in the spectra of the adiabatic populations. This indicates that the oscillations corresponding to these peaks have their origin in the mixing of local vibronic transitions by the Coulomb coupling.

This statement can be illustrated considering the exact population dynamics of the excited states of an electronic dimer without coupling to a bath. The dynamics of this system in both representations is given by the Schrödinger equation

$$H \left| \Psi(t) \right\rangle = i\frac{\partial}{\partial t} \left| \Psi(t) \right\rangle, \tag{3.3}$$

where the Hamiltonian in the diabatic basis is given by

$$H_{\text{dia}} = \begin{pmatrix} E_{e_1} & J \\ J & E_{e_2} \end{pmatrix} \tag{3.4}$$

and the corresponding adiabatic Hamiltonian by

$$H_{\text{adia}} = \begin{pmatrix} E_1 & 0 \\ 0 & E_2 \end{pmatrix}. \tag{3.5}$$

The adiabatic state energies are connected to the diabatic ones via Eq. (2.23). Expanding the time-dependent wave function $|\Psi(t)\rangle$ in terms of the diabatic and adiabatic states yields

$$|\Psi(t)\rangle_{\text{dia}} = \left(\begin{array}{c} A_{e_1}(t)\, e^{-iE_{e_1}t}\, |e_1\rangle \\ A_{e_2}(t)\, e^{-iE_{e_2}t}\, |e_2\rangle \end{array} \right) \tag{3.6}$$

and

$$|\Psi(t)\rangle_{\text{adia}} = \left(\begin{array}{c} A_1(t)\, e^{-iE_1t}\, |1\rangle \\ A_2(t)\, e^{-iE_2t}\, |2\rangle \end{array} \right), \tag{3.7}$$

respectively. The population of an arbitrary state is given by the absolute square of the corresponding expansion coefficients, i.e.

$$P_i = |A_i|^2 = A_i A_i^*. \tag{3.8}$$

Inserting the wave function and the Hamiltonian in Eq. (3.3) leads to equations of motions for the expansion coefficients A_i. In the adiabatic representation the corresponding equations are given by

$$\frac{\partial}{\partial t} A_i(t) = 0, \tag{3.9}$$

which yields immediately that the expansion coefficients and thus the populations of the adiabatic states are constant. However, due to the Coulomb coupling the equations of motion of the diabatic expansion coefficients, i.e.

$$\frac{\partial}{\partial t} A_{e_1}(t) = -iJ\, e^{i\omega_{e_1 e_2}t} A_{e_2}(t) \tag{3.10}$$

and

$$\frac{\partial}{\partial t} A_{e_2}(t) = -iJ\, e^{-i\omega_{e_1 e_2}t} A_{e_1}(t) \tag{3.11}$$

with $\omega_{e_1 e_2} \equiv E_{e_1} - E_{e_2}$, are coupled to each other. Solving the system of equations leads to

$$A_{e_2}(t) = b_1\, e^{i\Omega_1 t} + b_2\, e^{i\Omega_2 t}, \tag{3.12}$$

where the coefficients b_i need to be adjusted according to the initial conditions and the frequencies Ω_i are defined as

$$\Omega_{1,2} = \frac{\omega_{e_1 e_2}}{2} \pm \frac{1}{2}\sqrt{\omega_{e_1 e_2}^2 + 4J^2}. \tag{3.13}$$

Thus, the population of the state $|e_2\rangle$ reads

$$P_{e_2}(t) = A_{e_2}(t)A_{e_2}^*(t)$$
$$= |b_1|^2 + |b_2|^2 + b_1 b_2^* \, e^{i(\Omega_1 - \Omega_2)t} + b_1^* b_2 \, e^{-i(\Omega_1 - \Omega_2)t} \tag{3.14}$$

and the population oscillates with a frequency of $\Omega_1 - \Omega_2$, which corresponds to the energy gap between the adiabatic states, cf. Eq. (2.23).

The oscillation frequencies of the rightmost peaks in the FS of the diabatic populations for I and II, Fig. 3.3 panels a and b, coincide indeed very accurately with the gap energy of the adiabatic electronic states, whereas the corresponding peak for IV is shifted towards higher frequencies. This is a result of the mixing of the electronic and vibrational DOFs due to the strong vibronic coupling, which leads to a higher energy of the vibronic levels in comparison to the corresponding levels in the purely electronic model. In III the electronic peak in the FS of the diabatic population can barely be identified due to the large effect of the vibronic coupling, although there are some changes in the FS of the adiabatic population in comparison to the FS of the diabatic population. To summarize, one peak in the FS of the diabatic populations can be assigned to be of electronic origin. This peak is absent in the FS of the adiabatic populations due to the nature of the transformation between diabatic and adiabatic representation. Thus, in contrast to the diabatic populations, the adiabatic populations, Fig. 3.2, feature no oscillations of electronic origin.

However, the vibronic and the system-bath couplings lead not only to a relaxation onto energetically lower states, but also to additional oscillations even for a weak vibronic coupling (I and II) although this effect is more pronounced for a strong vibronic coupling (III and IV). The peaks in the FS, which have their origin in the vibronic coupling, show a stronger dependence on the bath coupling strength $\tilde{\gamma}$ than the electronic peaks. In particular, a larger peak broadening indicates a stronger damping of the corresponding oscillations with increasing $\tilde{\gamma}$ in comparison to their electronic counterparts.

In analogy to the electronic dimer, the key property to understand the population dynamics of the vibronic dimer and especially its oscillatory behaviour for arbitrary Huang-Rhys factors is the full adiabatic vibronic level structure, cf. app. C. Similar to the frequency of the electronic oscillations in the diabatic electronic representation, which corresponds to the energy gap of the adiabatic electronic states, the frequency of the vibronic oscillations in the diabatic vibronic picture corresponds to the energy gap between certain adiabatic vibronic levels. Due to the implicit treatment of the intra-molecular vibrations, the diabatic populations, displayed in Fig. 3.2, correspond to a diabatic vibronic representation and therefore feature the vibronic oscillations. Note that the electronic adiabatic populations shown in Fig. 3.2 are not adiabatic with respect to the vibronic basis and thus still feature vibronic oscillations, cf. app. C. The individual populations of the adiabatic and diabatic vibronic states are not available in the context of an implicit treatment of the intra-molecular vibrations as a heat bath. Nevertheless, the diabatic vibronic level scheme can be mimicked in the present system-bath model, as the properties of the intra-molecular modes, S and ω_{vib}, are known.

Introducing the diabatic vibronic states $|ab\mu\nu\rangle$, where a and b denote the electronic state and μ and ν the vibrational state of the first and second monomer, respectively, allows one to calculate the energy and oscillator strength. The latter is given by the normalized Boltzmann-weighted transition dipole square of the adiabatic vibronic states

$$|\alpha'\rangle = \sum_{ab\mu\nu} c_{ab\mu\nu} |ab\mu\nu\rangle, \qquad (3.15)$$

which can be obtained via numerical diagonalization of the diabatic vibronic Hamiltonian [79, 80]. The diabatic vibronic aggregate states represent either a vibronic excitation of the first monomer ($|eg\mu 0\rangle$), the second monomer ($|ge0\nu\rangle$), a vibrational excitation of the vibronically unexcited monomer ($|ge\mu\nu\rangle$ with $\mu > 0$ or $|eg\mu\nu\rangle$ with $\nu > 0$), or a vibrational excitation of the aggregate ground state ($|gg\mu\nu\rangle$). Furthermore, the character of the adiabatic vibronic aggregate states can be classified according to the overall number of vibrational excitations in the excitonically excited aggregate states, i.e.

$$\chi^{(0)} = c_{ge00}^2 + c_{eg00}^2 \qquad (3.16)$$

$$\chi^{(1)} = c_{ge10}^2 + c_{ge01}^2 + c_{eg10}^2 + c_{eg01}^2 \qquad (3.17)$$

$$\chi^{(2)} = c_{ge20}^2 + c_{ge02}^2 + c_{eg20}^2 + c_{eg02}^2 + c_{ge11}^2 + c_{eg11}^2 \qquad (3.18)$$

$$\vdots$$

The contribution of the states with zero vibrational excitations $\chi^{(0)}$ can be identified as the electronic contribution. Inspecting the properties of the adiabatic vibronic levels, Fig. 3.4, one can see that the distribution of the oscillator strength corresponds to the intensity distribution in the absorption spectra, Fig. 3.1.

For the weak vibronic coupling (I and II), levels with a large electronic contribution carry the major part of the oscillator strength, whereas for the strong vibronic coupling (III and IV) also levels with a large vibrational contribution gain significant oscillator strength. This leads to the appearance of vibronic peaks in the corresponding absorption spectra. Further, the peak broadening characteristics for III and IV, in particular that some peaks show a stronger broadening with increasing $\tilde{\gamma}$ than others, can be explained considering the character of the underlying adiabatic vibronic levels. For example, the peak at $(E - E_0)/\omega_{\text{vib}} \approx 3$ in Fig. 3.1 panel d, which corresponds to the 17th adiabatic vibronic level (Fig. 3.4 panel d), shows a larger peak broadening with increasing $\tilde{\gamma}$ than the peak at $(E - E_0)/\omega_{\text{vib}} \approx 2$, which corresponds to the eleventh level. This is correlated to the character of these levels, i.e. the eleventh level has a dominant electronic contribution, whereas the 17th level has a dominant $\chi^{(1)}$. According to the present system-bath model, in which the intra-molecular vibrations are coupled to the environmental DOFs, adiabatic vibronic aggregate states with a stronger vibrational excitation are stronger coupled to the environmental bath. Thus, states, which have a larger

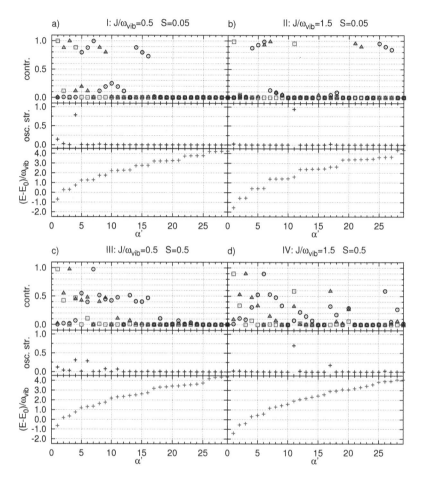

Figure 3.4.: Energy, oscillator strength, $\chi^{(0)}$ (red squares), $\chi^{(1)}$ (blue triangles) and $\chi^{(2)}$ (black circles) of the adiabatic vibronic states corresponding to the four dimer scenarios. The properties were obtained via direct diagonalization of a diabatic vibronic Hamiltonian incorporating ten vibrational levels for the ground and each excited monomeric state, respectively. The reference energy E_0 is defined as in Fig. 3.1.

$\chi^{(n>0)}$ in comparison to some other states, are stronger affected by the inter-bath coupling than the latter.

In order to interpret the oscillatory behaviour of the populations, it is necessary to select the contributing diabatic and the corresponding adiabatic vibronic states. Owing to the large number of adiabatic vibronic states and their, especially for strong vibronic couplings, dense energy spectrum, this task requires the implementation of certain rules. These can be developed inspecting the diabatic vibronic Hamiltonian, Eq. (C.4). First, at least one of the involved diabatic vibronic levels needs to be initially populated. Second, having in mind the analytical solution of the electronic model one notes that only coupled diabatic vibronic levels feature oscillations. The coupling between the diabatic vibronic levels is determined by the product of the Coulomb coupling strength J and the Franck-Condon overlaps of the vibrations in the monomeric ground and excited states, Eq. (C.6). Note that the Franck-Condon overlap depends crucially on the Huang-Rhys factor S. For small Huang-Rhys factors only the Franck-Condon overlaps of configurations, which differ in none or one vibrational quantum, have a pronounced amplitude. Therefore, only pairs of adiabatic vibronic aggregate states, of which one has a significant $\chi^{(v)}$ and the other has a significant $\chi^{(v)}$ or $\chi^{(v\pm1)}$, couple strongly. In contrast, the Franck-Condon overlaps of configurations, which differ in more than one vibrational quantum, are no longer negligible for larger Huang-Rhys factors. Thus, additional couplings become important and one needs to consider also pairs of states, of which one has a significant $\chi^{(v)}$ and the other has a significant $\chi^{(v\pm n)}$. The value of n depends crucially on S.

Due to the choice of the initial conditions for the investigation of the population dynamics of the four scenarios, i.e. a population of the second adiabatic electronic level, only vibronic states with a significant electronic contribution are initially populated. Following the rules introduced above it is possible to reveal the origin of the peaks in the FS. First, one needs to identify the levels with large oscillator strengths. Second, all energetically lower levels, whose energy gap to one of the identified levels matches the peak frequency need to be considered. Finally, one needs to compare the $\chi^{(v)}$ of the identified levels with the $\chi^{(v)}$ and $\chi^{(v\pm n)}$ of the considered energetically lower levels. Via the first two steps of the presented algorithm the first peak in the FS of the populations for I, Fig. 3.3 panels a and e, can be assigned to the energy difference between the fourth and second, and the fourth and third level, respectively. Whereas level four has a dominant electronic contribution the levels two and three are of dominantly vibronic character. Thus, the corresponding oscillations in the populations, cf. Fig. 3.2, are of vibronic origin, which is in agreement to the result obtained by a comparison between the adiabatic and diabatic FS. The same algorithm works for the first and second peak in the FS for II, which can be assigned to the energy gaps between the eleventh and the sixth, the eleventh and 21st and the eleventh and 22nd level, respectively. For III and IV, the adiabatic vibronic level structure is much more complex leading to a variety of superpositions, which yield the complex oscillatory behaviour of the populations.

The initial condition used so far, i.e. the population of the second adiabatic state, is rather different in comparison to an optical excitation of the system and thus the dynamics after absorption of light might differ significantly from the dynamics displayed in Fig. 3.2. The 2D-spectroscopy technique provides a powerful experimental tool to study the population dynamics after an optical excitation and the correlations between the different states [64, 65, 68]. However, the interpretation of the obtained 2D-spectra suffers from the high complexity of the investigated systems. Thus, comparison between experimental and calculated 2D-spectra provides not only a benchmark for the used theoretical model, but also facilitates the interpretation of the experimental spectra. The 2D-spectra for the different dimer scenarios were calculated using the response function formalism presented in Sec. 2.4 with the Rostock HEOM package.

Note that the calculation of 2D-spectra is computationally demanding even for small systems as the response functions, Eq. (2.157), need to be evaluated for various combinations of t_1, t_2 and t_3 to obtain sufficient data. The resolution of the calculated 2D-spectra depends crucially on the length of the time intervals $[t_1^{(\mathrm{min})}; t_1^{(\mathrm{max})}]$ and $[t_3^{(\mathrm{min})}; t_3^{(\mathrm{max})}]$ covered by the propagation. Additionally the time

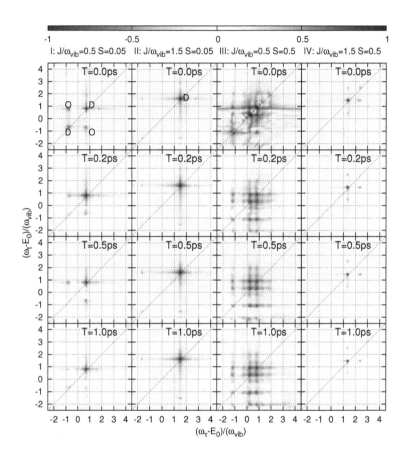

Figure 3.5.: 2D-spectra for the four dimer scenarios and weak inter-bath coupling $\tilde{\gamma} = 50$ cm^{-1} for $T = 0$ fs, $T = 200$ fs, $T = 500$ fs and $T = 1000$ fs. The spectra were normalized according to the maximum of the spectra for $T = 0$ fs. The diagonal and off-diagonal peaks for I and II are indicated by **D** and **O** in the upper panels, respectively

step between different $t_{1/3}$, i.e. the sampling of the corresponding intervals, needs to be accurate enough to cover all spectral features.

The 2D-spectra for the different dimer scenarios, Fig. 3.5 and Fig. 3.6, were calculated using a sampling of 0.5 fs and an interval

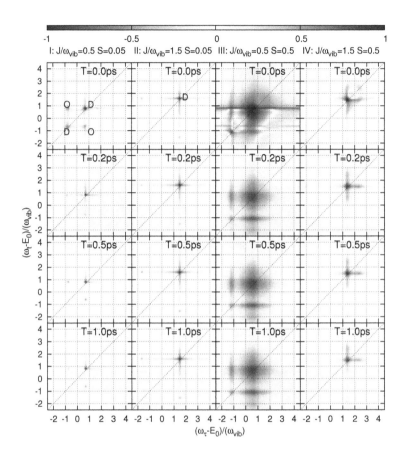

Figure 3.6.: Same as Fig. 3.5 for $\tilde{\gamma} = 200 \text{ cm}^{-1}$.

length of 2048 fs for both τ and t. According to this setup 4096^2, i.e. approximately 17 million, calculation steps are needed to calculate one 2D-spectrum. The resolution of the 2D-spectra is $\Delta\omega_{t/\tau} \approx 4 \text{ cm}^{-1}$, which is sufficient for an analysis with respect to possible oscillatory features of the peaks. However, due to the computational effort no sufficient resolution for T could be obtained during the present work. Furthermore, the response functions do not decay sufficiently during the interval length of 2048 fs, which leads to instabilities of the Fourier transform with respect to τ and t, Eq. (2.153). In comparison, the linear absorption spectra, Fig. 3.1, were obtained with a propagation time of 16 ps. Due to the instabilities of the Fourier transform, negative features arise in the 2D-spectra, which should be absent as the excited state absorption is not present in the current model. The convergence level of the hierarchy was decreased as well, i.e. $\mathcal{N} = 5$ instead of $\mathcal{N} = 9$, to reduce the overall computational effort.

The diagonals ($\omega_\tau = \omega_t$) of the 2D-spectra for the different scenarios for $T = 0$ fs resemble the linear absorption spectra, cf. Fig. 3.1. Similar to the linear absorption spectra, the 2D-spectra for I, II and IV have less features than the spectra for III. Again, the spectra for I and II are not so strongly affected by an increase of the inter-bath coupling constant $\tilde{\gamma}$ as their peaks correspond mainly to levels with dominant electronic character. The spectra for III show a variety of distinct cross-peaks, which indicate, similar to the FS of the populations, a coupling of the corresponding aggregate states, whereas the spectra for I, II and IV reveal only a few weak cross-peaks. Furthermore, the spectra for all scenarios show a rapid intensity redistribution within the first 200 fs but reveal in contrast to the population dynamics no significant changes for $T = 500$ fs and $T = 1000$ fs. This is due to the difference in the initial conditions. For the population dynamics simulation, Fig. 3.2, the initial condition was chosen such that only the second adiabatic electronic level was populated, whereas the interaction with a field excites the vibronic levels. To achieve a similar distribution of the initial populations and take into account the interactions of the system with the second and third pulse, one needs to propagate the system treating the system-field interaction explicitly, e.g., via the interaction operator H_{int}, Eq. (2.24). Note that this feature is not implemented in the Rostock HEOM package so far. Another difference between the population dynamics and the dynamics of the spectra arises due to the fact that the lower vibronic states, to which the population relaxes, carry only a small part of the oscillator strength according to the positive Coulomb coupling (H-aggregate). Therefore, the dynamics of these states is not observable.

To summarize this section, the population dynamics of dimer systems depends strongly on the interplay between the system and the bath. In particular the value of the spectral density at the transition energy between the adiabatic levels, which corresponds to the coupling strength of the transition to the bath, is crucial for the relaxation dynamics. Thus in the context of the used MBO model for a particular set of system variables, i.e. the Coulomb coupling strength J and the energy of the diabatic states E_{e_1} and E_{e_2}, an increase of the vibronic coupling strength (the Huang-Rhys factor S) will lead to a faster relaxation. Further, for a constant vibronic coupling and a fixed diabatic state configuration, the Coulomb coupling strength J determines the dynamics as it tunes the energy gap between the adiabatic states. The oscillatory features, which appear in the population dynamics, can be interpreted using the full adiabatic vibronic level structure of the system. However, this structure is not available for larger aggregates. Nevertheless, as a rule of thumb one can assign oscillations, which appear in the diabatic but not in the adiabatic electronic population, to be of electronic origin. The exact correlation between the population dynamics and the dynamics of the 2D-spectra will be in the focus of further investigations.

3.2. Funnel-like aggregate

In the previous section the population dynamics of a dimer system with a single component MBO spectral density was investigated. However, light-harvesting complexes usually consist of several molecules in a rather complex environment. The specific tasks of these specialized complexes in the overall photosynthetic process might vary. Whereas some complexes are specialized on the absorption of light and serve as light-harvesting antennae, e.g., the LH2 complex of purple bacteria, others, like the FMO complex of green sulphur bacteria, serve as energy funnels, which collect the energy of the antennae and transfer it to the reaction centre. There the energy is used for charge separation, which drives the chemical processes of photosynthesis [1]. To address the question of how the environment influences the population dynamics in a funnel-like aggregate, a model system consisting of eight monomers will be considered in the following. The monomers are grouped into four dimers, cf. Fig. 3.7 panel a. Note that such a dimerization is common in various light-harvesting complexes, e.g., the LH2 complex of purple bacteria [5, 6] and the FMO complex. The one-exciton part of the system Hamiltonian, cf. Eq. (2.6), is given by

$$H_{\mathrm{s}}^{(1)} = \hat{1}E_{\mathrm{c}} + \begin{pmatrix} \frac{3}{2}\Delta E & J_1 & 0 & 0 & 0 & 0 & 0 & 0 \\ J_1 & \frac{3}{2}\Delta E & J_2 & 0 & 0 & 0 & 0 & 0 \\ 0 & J_2 & \frac{1}{2}\Delta E & J_1 & 0 & 0 & 0 & 0 \\ 0 & 0 & J_1 & \frac{1}{2}\Delta E & J_2 & 0 & 0 & 0 \\ 0 & 0 & 0 & J_2 & -\frac{1}{2}\Delta E & J_1 & 0 & 0 \\ 0 & 0 & 0 & 0 & J_1 & -\frac{1}{2}\Delta E & J_2 & 0 \\ 0 & 0 & 0 & 0 & 0 & J_2 & -\frac{3}{2}\Delta E & J_1 \\ 0 & 0 & 0 & 0 & 0 & 0 & J_1 & -\frac{3}{2}\Delta E \end{pmatrix}$$

$$(3.19)$$

where E_{c} denotes the average of the diabatic excitation energies, the intra-dimer Coulomb coupling strength is denoted by J_1 and the inter-dimer Coulomb coupling strength by J_2. The diabatic electronic excitation energies are assumed to be the same within a dimer. In contrast the excitation energies of neighbouring dimers differ by some constant energy ΔE. Again four different scenarios are considered, see Tab. 3.2. The scenarios A and B cover the situation that the inter-dimer Coulomb coupling is equal to the intra-dimer Coulomb coupling, i.e. $J_1 = J_2$, whereas for C and D the intra-dimer Coulomb coupling is ten times larger than the inter-dimer Coulomb coupling, i.e. $J_1 = 10J_2$.

The adiabatic electronic level structure is characterized by the energy of the levels and the amplitude of the exciton states, that is the square of the expansion coefficient in Eq. (2.22). It features a

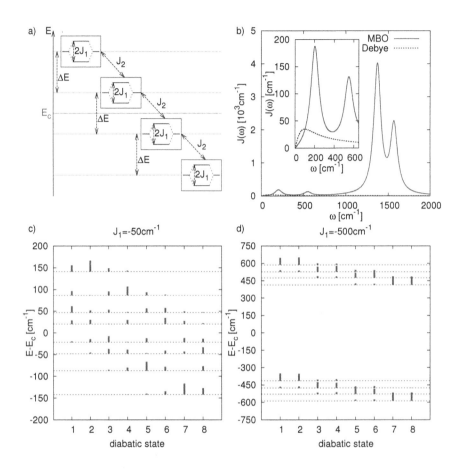

Figure 3.7.: (a) Schematic view of the funnel model system. (b) Spectral density models (parameters given in the text and Tab. 3.3). (c) Amplitudes of the exciton states at the different monomer sites for $J_1 = J_2$. (d) Same as (c) for $J_1 = 10J_2$

distinct separation into an upper and a lower manifold for C and D, Fig. 3.7 panel d. In contrast to this separation the gap energies between the levels for A and B are of the same order of magnitude, Fig. 3.7 panel c. The adiabatic levels of the upper/lower manifold

Table 3.2.: Intra-dimer Coulomb coupling strengths for various scenarios. The inter-dimer Coulomb coupling strength ($J_2 = -50$ cm^{-1}) and the gap energy ($\Delta E = 50$ cm^{-1}) are the same for all scenarios.

scenario	J_1 [cm^{-1}]	Spectral density model
A	-50	MBO
B	-50	Debye
C	-500	MBO
D	-500	Debye

Table 3.3.: Components of the MBO spectral density for the funnel model system.

component ξ	S_ξ	ω_ξ [cm^{-1}]	$\tilde{\gamma}_\xi$ [cm^{-1}]
1	0.197	206.0	100.0
2	0.215	211.0	100.0
3	0.037	552.0	100.0
4	0.208	1371.0	100.0
5	0.083	1570.0	100.0

for C and D stem from the upper/lower exciton states of the individual dimers. These states are localized on the associated dimers, i.e. the amplitude of the exciton states at the monomers which belong to the other dimers is small. In A and B the adiabatic states are delocalized over the whole aggregate. Further, two different spectral densities are considered for each system configuration, namely a Debye spectral density, Eq. (2.110) with $\eta = 0.7$ and $\tilde{\gamma} = 100$ cm^{-1} (B and D), and a five-mode MBO spectral density (A and C) which is specified in Tab. 3.3. Note that it is again assumed, that all monomers couple to individual but equal baths and thus the monomeric index m is skipped for the parameters of the spectral densities.

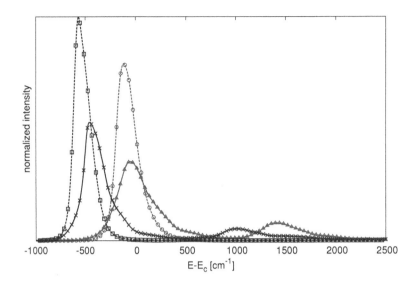

Figure 3.8.: Absorption spectra for the different scenarios for the funnel model system (A: red solid line and triangles; B: red dashed line and circles; C: blue solid line and crosses; D: blue dashed line and squares).

The MBO spectral density is adapted from the spectrum of the intra-molecular vibrations of a perylene bisimide pigment, which is a versatile building block for artificial aggregates used as molecular wires or antennae [76, 81–83]. Whereas the Debye spectral density has its maximum around 100 cm^{-1} and has no significant amplitude for frequencies higher than 600 cm^{-1}, the MBO spectral density has a significant amplitude up to 2000 cm^{-1}, Fig. 3.7 panel b.

Converged results for B and D were obtained with $\mathcal{N} = 6$ and $\mathcal{K} = 8$, i.e. one correlation function term per monomer. The results for the MBO scenarios (A and C), which were calculated with $\mathcal{N} = 4$ and $\mathcal{K} = 80$, that is two correlation function terms per mode and monomer, are not fully converged. However, the corresponding hierarchy size of approximately two million matrices, cf. Eq. (2.78), approaches the limit of the current version of the Rostock HEOM package, whereas the hierarchy size of approximately 3000 matrices for the Debye scenarios is far below that limit. This indicates the advantages of the Debye spectral density in the context of the HEOM formalism, but note that this spectral density is not suitable to model structured spectral densities. The error of the calculations for A and C, which can be estimated via comparison of the shown results with results obtained with $\mathcal{N} = 3$, is below ten percent.

The absorption spectra, Fig. 3.8, for B and D show only a single peak at the low energy side of the spectrum, which is specific for negative Coulomb couplings (J-aggregates). In contrast, the spectra for A and C feature in addition to the low-energy peak a vibronic progression due to the vibronic structure implied by the MBO spectral density, cf. Sec. 3.1. The stronger red-shift of the spectra for C and D with respect to the ones for A and B follows from the larger value of the intra-dimer Coulomb coupling J_1 in the former cases.

In order to study the spatial energy transfer within the aggregate the initial condition was chosen such that the highest adiabatic exciton state is populated. This state is mainly located at the first dimer in all scenarios, cf. Fig. 3.7. Note that the spatial energy transfer, that is the redistribution of the population among the diabatic (local) states, comes along with the relaxation between the adiabatic states.

The population dynamics of the dimers, Fig. 3.9, shows in all scenarios the same general behaviour, i.e. the population slides down in the energy funnel formed by the dimers towards the energetically lowest dimer. This process is faster for B and D in comparison to A and C. The difference between the population dynamics for A and B, as well as for C and D, respectively, is not that pronounced. However, the adiabatic populations, Fig. 3.10, show a significantly different behaviour for the different scenarios. For A and B the population of the states two to six increases to a similar value within a few tens of femtoseconds. This is due to the small energy gap between the states, e.g., approximately 200 cm^{-1} between state one and six. Both spectral densities, Fig. 3.7 panel b, feature a significant amplitude in this

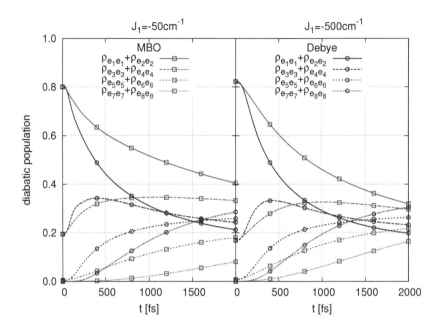

Figure 3.9.: Population dynamics of the dimers, cf. Fig. 3.7, calculated via the dynamics of the diabatic one-exciton states for the different scenarios (A: left panel red lines and squares; B: left panel blue lines and circles; C: right panel red lines and squares; D: right panel blue lines and circles).

frequency region and thus enable an effective relaxation, cf. Sec. 3.1. In contrast, for C and D, where the energy gap between the upper and lower manifold is approximately 800 cm^{-1}, the population is gradually transferred to the energetically lower states. For C this transfer occurs mainly in the lower exciton manifold after a fast relaxation of the population from state one to state five, whereas for D the population transfer occurs within the first few hundred femtoseconds in the upper and lower exciton manifolds. The different behaviour in both scenarios is due to the different amplitude of their spectral densities in the high-frequency region, i.e. the MBO spectral density features a significantly

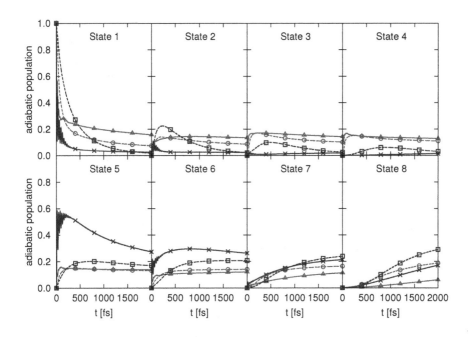

Figure 3.10.: Population dynamics of the adiabatic one-exciton states for the different scenarios (A: red solid lines and triangles; B: red dashed lines and circles; C: blue solid lines and crosses; D: blue dashed lines and squares). Note that the states are labelled in energetically decreasing order, cf. Fig. 3.7.

larger amplitude at the transition energy between state one and five than the Debye spectral density. Owing to the localization of these states on the same dimer, which leads to a strong coherence of these states for C, Fig. 3.7 panel d, there exists a pronounced oscillatory behaviour of the corresponding populations.

In order to investigate the influence of the different system and bath specifications on the exciton delocalization and the population dynamics, it is instructive to study not only the population of the states, i.e. the diagonal elements of the reduced density matrix, but also the off-diagonal elements of the reduced density matrix, the

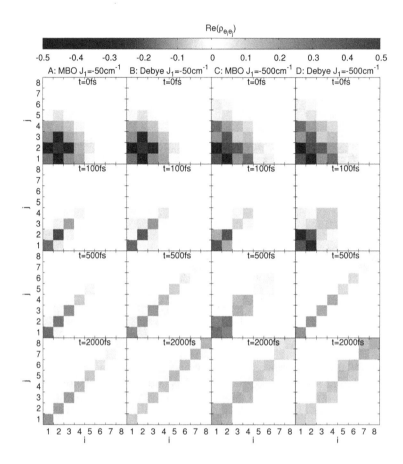

Figure 3.11.: Snapshots of the real part of the diabatic density matrix for various scenarios and propagation times.

so-called coherences. For all scenarios the initial diabatic density matrix, Fig. 3.11, is delocalized over several monomers according to the corresponding eigenvector decomposition, Fig. 3.7 panels c and d. Further, the coherences between the different monomers decay very fast within the first 100 fs. Whereas for A and B mainly coherences between neighbouring monomers survive, for C and D a strong coherence between the monomers which belong to the same dimer develops.

In contrast to A and C, for B and D additional coherences between monomers of different dimers remain significant. For long propagation times these additional coherences vanish and there appears a strict dimerization for C and D, i.e. there are only coherences between monomers which belong to the same dimer. This dimerization is not present for A and B, where only coherences between neighbouring monomers survive.

To summarize, the dynamics of the model aggregate depends, similarly to the dimer system, cf. Sec. 3.1, crucially on the interplay between the system and the bath. In particular, the shape of the spectral density is rather critical for the system dynamics, especially if there exist coupled levels with a large energy gap.

3.3. Fenna-Matthews-Olson complex

In the previous sections, the dependence of the system dynamics on the bath characteristics for a dimer model and a model of a funnel-like aggregate was investigated using model spectral densities. However, light-harvesting antennae like the Fenna-Matthews-Olson complex feature a more heterogeneous system structure as well as a complex spectral density. The FMO complex, which serves as an energy funnel linking the light-harvesting chlorosome with the reaction center in green sulphur bacteria [1], consists of three identical subunits. Each of the subunits is formed by an aggregate of seven *bacteriochloro-phyll a* (*BChl a*) molecules embedded in a protein environment [84], Fig. 3.12. Note that there exists an eighth *BChl a* molecule, which serves as a linker molecule to the chlorosome baseplate [85, 88, 89]. This additional molecule changes the weight of the energy pathways through the FMO complex and might suppress the observed coherent oscillations in the diabatic populations of sites one and two, Fig. 3.12, which are strongly coupled to each other [89]. However, the eighth *BChl a* molecule will not be considered in the present FMO model. For a detailed study of its influence on the dynamics of the FMO complex see Ref. [41].

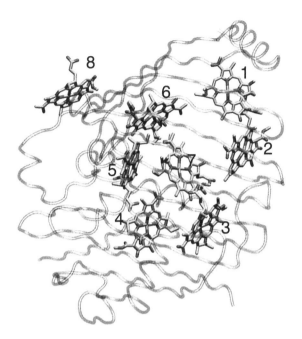

Figure 3.12.: Structure of the monomeric subunit of the FMO complex of *Prosthecochloris aestuarii* consisting of eight *BChl a* molecules (labelled) embedded in a protein environment of folded β-sheets (blue). The structural data was obtained via X-ray crystallography experiments [85] and is available via the RCSB protein data bank (PDB ID: 3EOJ) [86]. The figure was created using the VMD program [87].

Due to the structure of the monomeric subunit of the FMO complex and the influence of the protein environment, the Coulomb coupling between the *BChl a* molecules as well as the site energies are, in contrast to the model used in Sec. 3.2, rather heterogeneous. This leads to two primary energy transport pathways towards the *BChl a* molecule with the lowest site energy (site three), that is from site one via site two to site three and from site five via sites six and seven to site three. The diabatic one-exciton Hamiltonian, which corresponds to the labeling of the *BChl a* molecules in Fig. 3.12 and the structural data from *Prosthecochloris aestuarii*, is given (in cm^{-1}) by [89]

$$H_s^{(1)} = \hat{1}E_l + \begin{pmatrix} 310.0 & -97.9 & 5.5 & -5.8 & 6.7 & -12.1 & -10.3 \\ -97.9 & 230.0 & 30.1 & 7.3 & 2.0 & 11.5 & 4.8 \\ 5.5 & 30.1 & 0.0 & -58.8 & -1.5 & -9.6 & 4.7 \\ -5.8 & 7.3 & -58.8 & 180.0 & -64.9 & -17.4 & -64.4 \\ 6.7 & 2.0 & -1.5 & -64.9 & 405.0 & 89.0 & -6.4 \\ -12.1 & 11.5 & -9.6 & -17.4 & 89.0 & 320.0 & 31.7 \\ -10.3 & 4.8 & 4.7 & -64.4 & -6.4 & 31.7 & 270.0 \end{pmatrix}$$
$$(3.20)$$

with $E_l = 12195$ cm^{-1}.

Although there were attempts to calculate the spectral density of the FMO complex via molecular dynamics simulations [23], the most reliable available data stems from measurements of the spectral density via a combination of fluorescence line narrowing (FLN) and temperature-dependent linear absorption experiments [9]. Whereas the fluorescence line narrowing experiments provide access to the shape of the spectral density, the overall Huang-Rhys factor, i.e. the integral over the spectral density, cf. Eq. (2.114), needs to be evaluated via absorption measurements [9]. The FLN experiments were performed at a temperature of 4 Kelvin and the excitation was tuned such that only the pigment with the lowest energy has been excited. Therefore, the measured spectral density represents the monomeric spectral density of a *BChl a* molecule in the FMO complex, i.e. the couplings between the molecules are neglected. Further, the measurements do not account for the heterogeneity of the environment, i.e. it is assumed that all *BChl a* molecules within the complex have the same spectral density. However, recent calculations [23] show that there are differences between the spectral densities of the individual monomers. Note again that in contrast to Eq. (2.114), the experimentally obtained spectral density

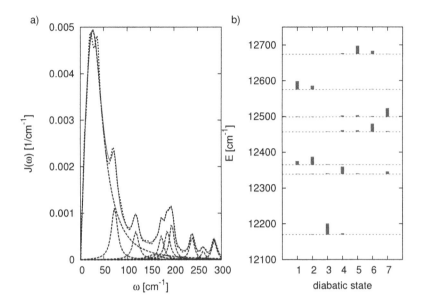

Figure 3.13.: (a) Spectral density of a *BChl a* molecule in the FMO complex (black dotted line: experimental data [9], red solid line: fit to MBO spectral density, blue dashed line: fit components). (b) Amplitudes of the exciton states at the different monomer sites, cf. Fig. 3.7.

is usually defined such that the overall coupling strength is given by

$$\tilde{S} = \int\limits_0^\infty d\omega\, J(\omega), \qquad (3.21)$$

that is both definitions differ by a prefactor of $\pi\omega^2$. Thus, to extract the parameters for quantum dynamics simulations, e.g., via a fit of the experimental data, one needs to adjust either the experimental data or the applied fit functions. For example the fit function corresponding to the MBO spectral density, cf. Eq. (2.117), is given by

$$J(\omega) = \sum_m \sum_\xi \frac{S_{\xi,m}}{\pi} \frac{\omega\omega_{\xi,m}\tilde{\gamma}_{\xi,m}}{(\omega^2 - \omega_{\xi,m}^2)^2 + \omega^2\tilde{\gamma}_{\xi,m}^2}. \qquad (3.22)$$

Table 3.4.: Components of the extracted MBO spectral density for a *BChl a* molecule in the FMO complex.

component ξ	S_ξ	ω_ξ [cm^{-1}]	$\tilde{\gamma}_\xi$ [cm^{-1}]
1	0.252	37.6	60.5
2	0.018	73.5	15.7
3	0.009	118.5	15.4
4	0.004	159.6	34.9
5	0.007	173.2	12.9
6	0.009	185.9	14.1
7	0.009	195.3	12.1
8	0.006	238.3	12.0
9	0.002	261.8	10.4
10	0.005	285.2	11.7

The measured spectral density of the *BChl a* molecules in the FMO complex, Fig. 3.13 panel a, was parametrized via a fit to a ten mode MBO spectral density. Note that the measured spectral density is normalized such that its integral corresponds to the Huang-Rhys factor of 0.45 [9]. Further, it is assumed that the spectral density is the same for all *BChl a* molecules in the complex and thus the monomer index m can be skipped. The obtained fit parameters are listed in Tab. 3.4. Note that the fit of the measured spectral density is quite accurate except for the frequency range below 50 cm^{-1} where the double-peak structure of the experimental data is approximated by a single MBO component. However, this approximation should be sufficient as the energy gaps between the adiabatic states, Fig. 3.13 panel b, which can be associated with the first main energy pathway, i.e. state six, three and one, are larger than 100 cm^{-1}.

To investigate the dynamics of the system corresponding to the first main energy pathway, namely from site one via site two to site three, the initial condition was chosen such that the first diabatic state is

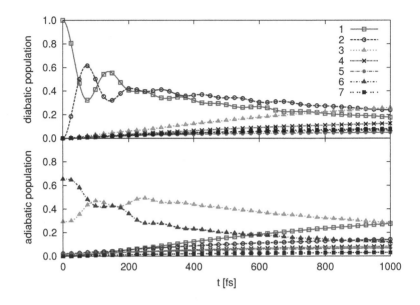

Figure 3.14.: Diabatic and adiabatic population dynamics of the FMO complex after an initial population of the first diabatic state.

populated. This corresponds to an initial population of the sixth and third adiabatic state, cf. Fig. 3.13 panel b. Note that the results are not converged for the given hierarchy setup ($\mathcal{N} = 3$, $\mathcal{K} = 140$) but a larger hierarchy cannot be treated with the current version of the Rostock HEOM package. Again, the error was estimated via comparison with a calculation with a lower convergence level. It is approximately twenty percent.

The dynamics of the diabatic population, Fig. 3.14, shows a pronounced oscillation between populations of the first and second diabatic state, whereas the population of the third diabatic state, as well as those of the states four to seven, shows an almost linear behaviour. This is in agreement to the population dynamics reported in the literature [89, 90]. The frequency of the oscillation, which is approximately 250 cm^{-1}, does not fit to the energy gap between the corresponding adiabatic states six and three, cf. Fig. 3.13 panel b, which is 210 cm^{-1}.

This indicates, in combination with the oscillations of the adiabatic population of states six and three, Fig. 3.14, that the oscillation is of vibronic origin, cf. Sec. 3.1. The small population of the diabatic states four to seven and the corresponding adiabatic states two, four, five and seven point towards the possibility to model the dynamics of the FMO complex by a reduced three-site model, which would reduce the numerical effort by far. A comparison of the full FMO model with such a reduced model, including the first, second and third site, is given in Ref. [41]. Note that the reduced model ansatz is valuable to study individual transfer pathways, but might by insufficient to study the dynamics after an optical excitation. The latter would lead to an initial excitation of various monomers and thus also pathways, which are not covered by a reduced model, might be important. As it was mentioned in Sec. 3.1, the connection of the oscillations in the population dynamics to the oscillations observed in 2D-spectroscopy experiments [16] is non-trivial and needs to be revealed by further investigations.

4. Summary

Since the observation of long lasting oscillations in the 2D-spectra of the FMO complex in 2007 [16], the influence of coherence on the exciton dynamics in natural photosynthetic antennae, which was long considered to be of hopping-like nature sufficiently described by rate equations, is of peculiar interest in the context of artificial photosynthesis. In order to study the dynamics, sophisticated methods, which are able to describe the dissipative quantum dynamics of molecular aggregates embedded in a protein environment rather accurately, are required. The HEOM formalism, outlined in Sec. 2.2.2 and implemented, e.g., in the Rostock HEOM package, provides such a powerful tool. Here, the influence of the environment is mimicked via an in principle infinite hierarchy of auxiliary density matrices. This leads to a considerable numerical effort even for small systems, which restricts the applicability of the HEOM method. However, recent improvements, concerning the method itself as well as the computational algorithms, facilitate the treatment of the dynamics of light-harvesting complexes.

In the present master thesis, the exciton dynamics of a dimer system, a model aggregate resembling light-harvesting antennae and the FMO complex were investigated using the HEOM method. The population dynamics, as well as the linear and 2D-spectra, of the dimer shows a strong dependence on the system and bath properties. In particular, the value of the spectral density at the transition energies of the system is crucial for the relaxation dynamics, as it reflects the coupling of the corresponding transition to bath. The oscillatory features, which appear in the population dynamics, can be interpreted using the adiabatic vibronic level structure of the dimer. Unfortunately, this structure is not available for larger aggregates. However, the origin of the oscillatory features can be determined by a comparison of the adiabatic and diabatic electronic populations. Oscillations, which stem from a superposition of vibronic levels, are present in both, whereas ones of electronic origin are only present in the latter. The connection of the oscillations in the populations to the ones in 2D-spectra needs to be investigated in more detail in the future.

Similar to the dimer system, the dynamics of the model aggregate is determined by the interplay of the system and the bath. Especially the weight of the transfer pathways within the aggregate depends crucially on the shape of the spectral density. The dynamics of the FMO complex, which was calculated using experimental data for the spectral density, indicates that coherences might play an important role in some of the transfer steps. In particular, the transfer step from the first to the second site seems to be governed by a coherence, which can be assigned by a comparison of the diabatic and adiabatic populations to be of vibronic origin.

A. The Rostock HEOM package

This appendix describes the Rostock HEOM package which was developed during the present master thesis. The package is available upon request. First, an overview of the package itself and some features of the implementation will be given. Second, the installation and the usage of the main program and some analysis tools will be described. Finally, the appendix contains the input documentation of the package.

A.1. Overview

The Rostock HEOM package is written in FORTRAN 90/95 and the subroutines are organized in several modules, Tab. A.1, corresponding to their function. In general the program execution consists of four parts, Fig. A.1. First, the input file is read by the program. Second, the system and bath parameters are processed, i.e. all operators are transformed to the eigenbasis of the Hamiltonian, the correlation function is calculated for the given spectral density, and the hierarchy is constructed. Third, the propagation corresponding to the defined task, i.e. the calculation of the population dynamics, the linear or 2D-spectra, is performed in the excitonic eigenbasis. Finally, the calculated values and a summary are printed to the output files. Note that some values are continuously printed, i.e. the population of the states or the autocorrelation function. The modules are designed to be independent of each other, i.e. the data is in general passed via global variables. Exceptions from this concept are made for the propagation routines.

The bath correlation function needs to be parametrized to fulfill the requirements of the HEOM formalism, cf. Sec. 2.2.2. So far the pack-

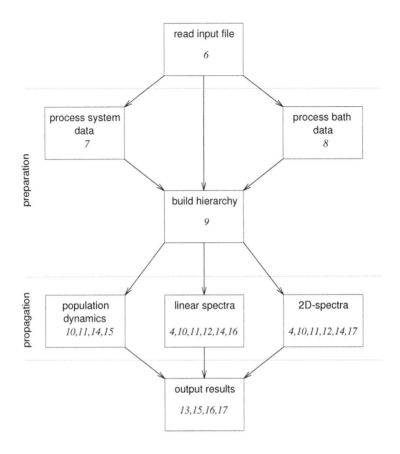

Figure A.1.: General structure of the current implementation. The italic numbers refer to the main modules, which are involved in the current step, cf. Tab. A.1.

age is able to process the MBO, Debye and Ohmic spectral density, cf. Eq. (2.117), Eq. (2.110) and Eq. (2.109). Whereas the corresponding correlation functions for the MBO and the Debye model are given by Eq. (2.119) and Eq. (2.111), respectively, the one corresponding to the Ohmic spectral density needs to be evaluated, e.g., via the Meier-Tannor parametrization scheme, cf. Eq. (2.126). Note that the

Table A.1.: List of modules (as of 2014-06-30).

module	No.	description
types_and_constants.f90	1	Data type definition, conversion factors and physical constants.
profiling.f90	2	Routines for a basic run time profiling of the code.
speed.f90	3	Routines to log the time needed for the individual propagation steps.
fft.f90	4	Interface to the FFTW3 library routines.
global_variables.f90	5	Declaration of global variables.
read_input_sections.f90	6	Routines to process the input file.
eigensystem.f90	7	Routines to calculate the eigensystem of the Hamiltonian via LAPACK routines and to transform all matrices and operators to the eigenbasis of the Hamiltonian.

poles of the Bose-Einstein distribution function, cf. Eq. (2.106), are treated via the Matsubara scheme. All correlation functions obey the required general form

$$C(t) = \sum_{k=1}^{\mathcal{K}} c_k \, e^{-\gamma_k t} + \delta C. \qquad (A.1)$$

An explicit treatment of many expansion terms results in a huge hierarchy which is not desirable at all. Therefore, only a small number of expansion terms, which needs to be specified in the input, will be treated explicitly. However, the residual part of the correlation func-

Table A.2.: Continued list of modules (as of 2014-06-30).

module	No.	description
correlation_function.f90	8	Routines to calculate the correlation function expansion into a sum of exponentials for the MBO Debye and Ohmic spectral density.
build_hierarchy.f90	9	Routines to build the required hierarchy, i.e. label the ADM, build the references tables and initialize the matrices.
rhs_heom.f90	10	Routine which evaluates the right-hand side of the HEOM.
ontheflyfiltering.f90	11	Routines for a numerical filtering of the hierarchy to reduce the computational effort.
spectra.f90	12	Routines to calculate the linear and nonlinear spectra.
summary.f90	13	Routines to create the log file.
propagation_algorithms.f90	14	Propagation routines.
population_dynamic.f90	15	Execution routines to calculate the population dynamics.
linear_signal.f90	16	Execution routines to calculate the linear spectrum.
nonlinear_signal.f90	17	Execution routines to calculate the nonlinear spectrum.

tion can be approximated following the method outlined in Ref. [36], i.e.

$$\delta C = \sum_{k=\mathcal{K}+1}^{\tilde{\mathcal{K}}} c_k\, \mathrm{e}^{-\gamma_k t} + \delta C', \qquad (A.2)$$

employing all terms up to a specific number $\tilde{\mathcal{K}}$. This approximation yields for the coefficient Δ, cf. Eq. (2.82),

$$\Delta = \sum_{k=\mathcal{K}+1}^{\tilde{\mathcal{K}}} \frac{c_k}{\gamma_k}. \qquad (A.3)$$

Note that the coefficient Δ is determined for each correlation function $C_\zeta(t)$, if there exist multiple environments connected to the same system.

The process of labeling the ADM (RDM) is divided into two parts. First, the algorithm follows the $\mathcal{K}$-dimensional quasi-tree structure corresponding to the hierarchy in pre-order setup (root-left-right labeling). The quasi-tree will be set up recursively and direct knot - leaf references will be created immediately. However, following the algorithm there will be some references, which are not created directly. These references will be added in a second step, which is the actual bottleneck of this algorithm. The labeling scheme is sketched in Fig. A.2.

The propagation algorithms, i.e. the Runge-Kutta 4 and the Runge-Kutta-Fehlberg 4/5 algorithm, require multiple copies of the ADM array to perform one propagation step. Therefore, beside the larger amount of time needed to propagate a larger hierarchy, the available memory limits the hierarchy size that can be treated with the current version of the package. The propagation routines are parallelized using two different schemes. There exists a shared parallelization via OpenMP, which is suited, e.g., for machines with rather low memory, and a parallel memory parallelization via MPI. The parallelization type can be chosen in the "make" file during the installation of the package, cf. App. A.2. As the calculation of 2D-spectra via the response function formalism requires a large number of propagations there exists also a parallel memory parallelization for the corresponding routine.

The hierarchy is automatically numerically filtered by a on-the-fly truncation algorithm, if the keyword *FILTERING* is given in the input. In the present implementation, all ADM which have no

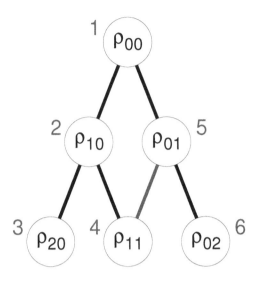

Figure A.2.: Scheme of ADM labeling system for $\mathcal{N} = 2$ and $\mathcal{K} = 2$. The labels of the ADM (red numbers) as well as the direct references (black lines) are created via running trough the quasi-tree corresponding to the hierarchy in pre-order setup. The indirect references which do not appear in the quasi-tree (red lines) are created by an additional routine.

element which is bigger than 10^{-8} are deleted from the hierarchy and the corresponding references are eliminated. The truncation routine is executed after every hundredth integration step.

A.2. Installation and usage of the programs

Please note that it is necessary to edit the compiler and library setup in the "make" file to adjust the program to the local environment. Note that some of the required libraries, that are FFTW3, LAPACK, and BLAS, might be included in more sophisticated libraries like the Intel Math Kernel Library (MKL).

The program makes use of both parallel and shared memory parallelization. Shared memory parallelization via OpenMP is used for the integrator routines and will be enabled by a compiler flag (e.g. -openmp for ifort). Parallel memory parallelization might either be used for parallelization of the integrator routines (compiler flag -Dmpi_int) or the 2D-spectra calculation (-Dmpi_2D). Note that mpi parallelization will be in general controlled by a pre-compiler (e.g. -fpp for the ifort compiler). To compile the programs it is recommended to execute the installation script *install.sh*. This script will create the executables. It is possible to add the executables to your shell environment via passing the location and name of your shell configuration file to the script, e.g.

$$./install.sh \ /.bashrc$$

The main HEOM program needs to be called with an input file as argument.

$$HEOM \ inputfile$$
$$or$$
$$mpirun \ -np \ N\_proc \ HEOM \ inputfile$$

For a detailed description of the requirements of the input file see App. A.3.

Two analysis programs are provided to calculate linear and nonlinear spectra, respectively. They need to be called with several arguments which are listed below.

$$linspecHEOM \ inputfile \ outputfile \ numberofsteps \ stepwidth$$
$$leftoutputborder \ rightoutputborder \ unitsystem \ abs/emi$$

$$2DspecHEOM \ inputfile \ outputfile \ xsteps \ ysteps \ xstepwidth \ ystepwidth$$
$$xdirection \ ydirection \ leftoutputborder \ rightoutputborder \ unitsystem$$

As unit systems so far eV, cm^{-1} and nm are implemented. *abs/emi* refers to absorption and emission for the linear spectrum (note that the latter is not implemented so far), $x - /y - direction$ indicates the directions of the underlying Fourier transform. The step width needs to be given in fs. Note that there exist two utility scripts *2Dspec.sh* and *joindat.sh* which facilitate the calculation of the 2D-spectra.

A.3. Input documentation

A.3.1. Run section

The run section contains all general information needed for the propagation, like the dimensionality of the reduced density matrix and all operators, the integration scheme and so on, as well as keywords to enable features like the on-the-fly filtering of the hierarchy. It is specified by

run-section
keywords (each in a single line) #comments
end-hierarchy-section

The keywords are listed below.

Table A.3.: List of run-section keywords (as of 2014-06-30)

keyword	description
$NSTATE = N$	required; specifies the dimensionality of the reduced density matrices and the operators, the value N needs to be a positive integer number
$NSTEP = N$	required; specifies the number of output steps (which may differ from the number of integrator steps), the value N needs to be a positive integer number; NOT required for 2D-spectra calculations

Table A.4.: Continued list of run-section keywords (as of 2014-06-30)

keyword	description
$TOUT = R$	required; specifies the output time step in fs, the value R needs to be a positive real valued number and an integer multiple of *tint*; NOT required for 2D-spectra calculations
$DEBUG$	optional; if this keyword is given, the list of references and the RDM labels will be written to separate log files
$THREADS = N$	optional; specifies the number of threads used for parallelization of the integration scheme, the value N needs to be a positive integer number; note that to avoid overload the number of threads will be adjusted by the program, nevertheless the value of N is the maximum number of possible threads used by the program
$TEMPERATURE = R$	required; specifies the temperature in K of the system, the value R needs to be a positive real valued number
$TRUNCORDER = N$	required; specifies the depth of the hierarchy, the value N needs to be a positive integer number
$DELTAAPPROX$	optional; enables the approximation of the residual part of the bath correlation function by a δ-function, cf. Eq. (2.82)

Table A.5.: Continued list of run-section keywords (as of 2014-06-30)

keyword	description
$FILTERING$	optional, not available for 2D-spectra calculations; enables the on the fly numerical filtering algorithm which deletes all reduced density matrices which contains only elements that are smaller than 10^{-8} from the hierarchy (compare [35, 36]); the adjustment routine is called after each 100 integrator steps
$ADIABATICINITIALRDM$	optional; declares that the systems reduced density matrix given in the initial RDM section is already represented in the adiabatic basis
$INTEGRATOR = C; R1; R2$	required,required,optional; specifies the integrator used to solve the HEOMs; the first argument denotes the integration scheme, the second one the initial integration step size and the last one the error tolerance for adaptive schemes, Runge-Kutta 4 ($C = RK4$) and the Runge-Kutta-Fehlberg 4/5 ($C = RKF4$) schemes are implemented

Table A.6.: Continued list of run-section keywords (as of 2014-06-30)

keyword	description
$SPECTRARANGE = R1; R2; C$	optional; specifies the range and unit system for spectra output; $R1$ denotes the left plot border, $R2$ the right plot border and C (cm-1, nm, eV) the unit system, default are 0;50000;cm-1
$ADIABATICCOUPLING$	optional; declares that the system-bath interaction operators given in the interaction section are represented in the adiabatic basis, which means that the bath couples to the adiabatic states directly

A.3.2. Task section

The tasks section contains the information which tasks shall be performed by the program. It is specified as follows.

task-section
keywords (each in a single line)
end-tasks-section

The keywords are listed below.

Table A.7.: List of task-section keywords (as of 2014-06-30)

keyword	description
population	outputs the population of the diabatic and adiabatic states to the file *pop.dat*

Table A.8.: Continued list of task-section keywords (as of 2014-06-30)

keyword	description
autocorrelation	outputs the autocorrelation function to the file *auto.dat*; note that a dipole operator is required to perform this task
absorptionspectrum	sets autocorrelation and calculates the linear absorption spectrum with the *linspecHEOM* algorithm; output will be written to *spec.dat*; note that a dipole operator and a specified plot-section are required
emissionspectrum	sets autocorrelation and calculates the linear emission spectrum with the *linspecHEOM* algorithm; output will be written to *spec.dat*; note that a dipole operator and a specified plot-section are required
2Dspectrum	calculates third order response functions (see Ishizaki and Tanimura, J. Chem. Phys. 125, 084501 (2006)) which are required to calculate the 2D-spectrum, a dipole operator as well as the 2D section are required.
2Dspectrumdetail	calculates the individual components of the third order response functions (see Chen et al. J. Chem. Phys. 132, 024505 (2010), eq. 15) which are required to calculate the 2D-spectrum, a dipole operator as well as the 2D section are required.

A.3.3. Parameter section

The parameter section contains all parameter definitions which might be used in the Hamiltonian, dipole, interaction, bath and initial-RDM section. Each parameter should be written in a single line. The section is specified as follows.

parameter-section
identifier=value:unit #comments
end-parameter-section

The identifiers must not be longer than 16 characters. So far as units electron volts (eV) and wave numbers (cm^{-1}) are supported. Note that all values without a unit (skip the colon) will be handled as they were in atomic units.

A.3.4. Hamiltonian section

The Hamiltonian section defines the system Hamiltonian in diabatic representation. Each element of the Hamiltonian should be written in a single line. The section is specified as follows.

hamiltonian-section
identifier [space] i [space] j [space] value/parameter
end-hamiltonian-section

There are two possible identifiers. S denotes symmetric matrix elements, that means $M_{ij} = M_{ji} = value$, and A denotes just the matrix elements M_{ij}. The value can be specified either directly or via a parameter identifier specified in the parameter section. All lines which do not start with one of the identifiers will be ignored and can be seen as comments. Note that all matrix elements which are not specified will be zero.

A.3.5. Dipole section

The dipole section defines the dipole operator in diabatic representation. Each element of the dipole operator should be written in a single line. The section is specified as follows.

dipole-section
identifier [space] i [space] j [space] value/parameter
end-dipole-section

There are two possible identifiers. S denotes again symmetric matrix elements of the type $M_{ij} = M_{ji} = value$ and A denotes just the matrix elements M_{ij}. The value can be specified either directly or via a parameter identifier specified in the parameter section. All lines which do not start with one of the identifiers will be ignored and can be seen as comments. Note that all matrix elements which are not specified will be explicitly zero.

A.3.6. Interaction section

The interaction section defines the system-bath interaction operators Q. Each element of a specific operator should be written in a single line. The section is specified as follows.

interaction-section
usedens=N
identifier [space] i [space] j [space] value/parameter
end-interaction-section

You can specify an arbitrary number of operators. Note that each operator will be connected to a spectral density given in the bath section by the keyword $usedens = N$. N refers to the number of a defined spectral densities. It is possible to connect multiple operators to one spectral density, e.g., couple multiple monomers to the same kind of bath. There are two possible identifiers for the matrix elements. S denotes symmetric matrix elements $M_{ij} = M_{ji} = value$ and A denotes just the matrix elements M_{ij}. The value can be specified either directly or via a parameter identifier specified in the parameter section. All lines which do not start with one of the identifiers will be ignored and can be seen as comments. Note that all matrix elements which are not specified will be explicitly zero.

A.3.7. Bath section

The bath section contains the definitions of the spectral densities. Each spectral density component should be written in a single line. The section is specified as follows.

bath-section
identifier [space] parameter list [space] N
end-bath-section

The identifier might either be *debye, ohmic* or *brownian*. The specific parameter lists are explained below, cf. Sec. 2.3 for the definition of the individual parameters. Note that you can use parameters defined in the parameter section. N defines the number of frequencies which will be explicitly handled within the HEOM formalism.

Table A.9.: List of supported spectral densities with parameters (as of 2014-06-30)

identifier	parameter list
debye	η_m [space] $\omega_{c,m}$ [space] N
ohmic	η_m [space] $\omega_{c,m}$ [space] N
brownian	$2S_{\xi,m}$ [space] $\omega_{c,m}^{(\xi)}$ [space] $\omega_{\xi,m}$ [space] N

A.3.8. Initial-RDM section

The initial-RDM section defines the initial top level reduced density matrix in diabatic representation. Each component of the initial RDM should be written in a single line. The section is specified as follows.

initial-RDM-section
identifier [space] i [space] j [space] value/parameter
end-initial-RDM-section

The initial top level RDM is given explicitly (all other RDMs will be set to zero). There are two possible identifiers for the matrix specification. S denotes symmetric matrix elements $M_{ij} = M_{ji} = value$ and A denotes just the matrix elements M_{ij}. The value can be specified either directly or via a parameter identifier specified in the parameter section. All lines which do not start with one of the identifiers will be ignored and can be seen as comments. Note that all matrix elements which are not specified will be explicitly zero.

A.3.9. 2D section

The 2D section defines the propagation times for the 3rd order response functions. It is specified as follows.

2D-section
t1=t1_start;t1_stepwidth;number_of_t1_steps
t2=t2_start;t2_stepwidth;number_of_t2_steps
t3=t3_start;t3_stepwidth;number_of_t3_steps
end-2D-section

The _start and _stepwidth values need to be defined in fs. The number of steps is an integer value.

B. Example inputs

This appendix provides example inputs for the Rostock HEOM package, cf. App. A, for the calculations presented in Sec. 3.1. Note that the inputs are written for the current version of the Rostock HEOM package (as of 2014-06-30), upwards compatibility with future versions is not granted. The numbers in front of the input keywords denotes the line numbers of the input file.

B.1. Population dynamics dimer system

1 run-section
2 TEMPERATURE=300.0
3 TRUNCORDER=9
4 NSTATE=2
5 TOUT=1.0
6 NSTEP=4000
7 INTEGRATOR=RKF4;0.05;1.0e-5
8 THREADS=8
9 FILTERING
10 DELTAAPPROX
11 ADIABATICINITIALRDM
12 end-run-section
13
14 task-section
15 population
16 end-task-section
17
18 parameter-section
19 E1=10000.0:cm-1
20 E2=10500.0:cm-1
21 J12=250.0:cm-1
22 eta=0.1 # Note eta=2S
23 gamma=200.0:cm-1
24 omega=500.0:cm-1
25 end-parameter-section
26
27 hamiltonian-section
28 S 1 1 E1
29 S 2 2 E2
30 S 1 2 J12
31 end-hamiltonian-section
32
33 bath-section

34 brownian eta gamma omega 2
35 end-bath-section
36
37 interaction-section
38 usedens=1
39 S 1 1 1.0
40 usedens=1
41 S 2 2 1.0
42 end-interaction-section
43
44 initialRDM-section
45 S 2 2 1.0
46 end-initialRDM-section

B.2. Linear absorption spectrum dimer system

1 run-section
2 TEMPERATURE=300.0
3 TRUNCORDER=9
4 NSTATE=3
5 TOUT=0.25
6 NSTEP=66000
7 INTEGRATOR=RK4;0.05
8 THREADS=8
9 FILTERING
10 DELTAAPPROX
11 SPECTRARANGE=5000.0;15000.0;cm-1
12 end-run-section
13
14 task-section
15 absorptionspectrum
16 end-task-section
17
18 parameter-section
19 E0=0.0:cm-1
20 E1=10000.0:cm-1
21 E2=10500.0:cm-1
22 J12=250.0:cm-1
23 eta=0.1
24 gamma=200.0:cm-1
25 omega=500.0:cm-1
26 end-parameter-section
27
28 hamiltonian-section
29 S 2 2 E1
30 S 3 3 E2
31 S 2 3 J12
32 end-hamiltonian-section
33

34 bath-section
35 brownian eta gamma omega 2
36 end-bath-section
37
38 interaction-section
39 usedens=1
40 S 2 2 1.0
41 usedens=1
42 S 3 3 1.0
43 end-interaction-section
44
45 dipole-section
46 S 1 2 1.0
47 S 1 3 1.0
48 end-dipole-section
49
50 initialRDM-section
51 S 1 1 1.0
52 end-initialRDM-section

B.3. 2D-spectrum dimer system

1 run-section
2 TEMPERATURE=300.0
3 TRUNCORDER=5
4 NSTATE=3
5 INTEGRATOR=RK4;0.05
6 THREADS=1
7 DELTAAPPROX
8 FILTERING
9 end-run-section
10
11 parameter-section
12 E0=0.0:cm-1
13 E1=10000.0:cm-1
14 E2=10500.0:cm-1
15 J12=250.0:cm-1
16 eta=0.1
17 gamma=200.0:cm-1
18 omega=500.0:cm-1
19 end-parameter-section
20
21 task-section
22 2Dspectrum
23 end-task-section
24
25 bath-section
26 brownian eta gamma omega 2
27 end-bath-section
28
29 interaction-section
30 usedens=1
31 S 2 2 1.0
32 usedens=1
33 S 3 3 1.0

34 end-interaction-section
35
36 hamiltonian-section
37 S 1 1 E0
38 S 2 2 E1
39 S 3 3 E2
40 S 2 3 J12
41 end-hamiltonian-section
42
43 dipole-section
44 S 1 2 1.0
45 S 1 3 1.0
46 end-dipole-section
47
48 initialRDM-section
49 S 1 1 1.0
50 end-initialRDM-section
51
52 2D-section
53 t1=0.0;0.5;1024
54 t2=0.0;25.0;41
55 t3=0.0;0.5;1024
56 end-2D-section

C. Electronic vs. vibronic basis

This appendix focusses on the connection of the electronic and vibronic basis in the adiabatic and diabatic representation for a dimer system. The one-exciton Hamiltonian of a dimer system in the electronic diabatic representation is given by, cf. Eq. (2.6),

$$H_{\text{dia}}^{(1,\text{el})} = \begin{pmatrix} E_{e_1} & J_{12} \\ J_{21} & E_{e_2} \end{pmatrix}, \tag{C.1}$$

with $J_{12} = J_{21} = J$. Diagonalization of this Hamiltonian yields the adiabatic electronic Hamiltonian

$$H_{\text{adia}}^{(1,\text{el})} = \begin{pmatrix} E_1 & 0 \\ 0 & E_2 \end{pmatrix}, \tag{C.2}$$

where the adiabatic energies E_1 and E_2 are given by Eq. (2.23). The relation between the adiabatic and diabatic electronic Hamiltonian can mathematically expressed as

$$H_{\text{adia}}^{(1,\text{el})} = S_{\text{el}}^{-1} H_{\text{dia}}^{(1,\text{el})} S_{\text{el}}. \tag{C.3}$$

Here the transformation matrix S_{el} is given by the eigenvectors of the diabatic Hamiltonian. Their elements can be identified as the expansion coefficients $c_{\alpha,m}^{(a)}$ in Eq. (2.22).

The diabatic vibronic Hamiltonian including the zeroth and first vibrational level at each site, cf. Sec. 3.1, is given by

$$
H_{\mathrm{dia}}^{(1,\mathrm{vib})} =
\left(
\begin{array}{cccc|cccc}
E_{e_1 00} & JF_{00}^{00} & 0 & 0 & JF_{00}^{10} & JF_{00}^{01} & 0 & JF_{00}^{11} \\
JF_{00}^{00} & E_{e_2 00} & JF_{00}^{10} & JF_{00}^{01} & 0 & 0 & JF_{00}^{11} & 0 \\
\hline
0 & JF_{10}^{00} & E_{e_1 10} & 0 & JF_{10}^{10} & JF_{10}^{01} & 0 & JF_{10}^{11} \\
0 & JF_{01}^{00} & 0 & E_{e_1 01} & JF_{01}^{10} & JF_{01}^{01} & 0 & JF_{01}^{11} \\
JF_{10}^{00} & 0 & JF_{10}^{10} & JF_{10}^{01} & E_{e_2 10} & 0 & JF_{10}^{11} & 0 \\
JF_{01}^{00} & 0 & JF_{01}^{10} & JF_{01}^{01} & 0 & E_{e_2 01} & JF_{01}^{11} & 0 \\
0 & JF_{11}^{00} & 0 & 0 & JF_{11}^{10} & JF_{11}^{01} & E_{e_1 11} & JF_{11}^{11} \\
JF_{11}00 & 0 & JF_{11}^{10} & JF_{11}^{01} & 0 & 0 & JF_{11}^{11} & E_{e_2 11}
\end{array}
\right)
, \quad (C.4)
$$

where $E_{a\mu\nu}$ denotes the energy corresponding to the diabatic aggregate state a with the vibrational excitation μ/ν in the first/second monomer and J the Coulomb coupling matrix element. $F_{\mu_a \nu_b}^{\mu'_{a'} \nu'_{b'}}$ represents the vibrational overlap (the Franck-Condon factors) between the vibrational configurations of the different aggregate states [79, 80]. The overall coupling elements are given by

$$
\langle m_a n_b \mu_a \nu_b | H_{\mathrm{dia}}^{(1,\mathrm{vib})} | m_{a'} n_{b'} \mu'_{a'} \nu'_{b'} \rangle = J(m_a n_b, m_{a'} n_{b'}) \langle \mu_a | \mu'_{a'} \rangle \langle \nu_b | \nu'_{b'} \rangle
$$
$$
= J F_{\mu_a \nu_b}^{\mu'_{a'} \nu'_{b'}}. \quad (C.5)
$$

Note that the Franck-Condon overlap integral, which is given for a single vibrational mode by

$$
\langle \mu_g | \mu'_e \rangle = e^{-S} \sum_{M=0}^{\mu_g} \sum_{N=0}^{\mu'_e} \frac{(-1)^N \sqrt{2S}^{M+N}}{M!N!}
$$
$$
\times \sqrt{\frac{\mu_g! \mu'_e!}{(\mu_g - M)!(\mu'_e - N)!}} \delta_{\mu_g - M, \mu'_e - N}, \quad (C.6)
$$

depends on the Huang-Rhys factor S. Similar to the diabatic electronic Hamiltonian, the diabatic vibronic Hamiltonian can be diagonalized introducing the transformation matrix S_{vib} yielding the adiabatic vibronic Hamiltonian, i.e.

$$H_{\text{adia}}^{(1,\text{vib})} = S_{\text{vib}}^{-1} H_{\text{dia}}^{(1,\text{vib})} S_{\text{vib}}. \tag{C.7}$$

However, for small Huang-Rhys factors ($S \leq 0.05$), i.e. the Franck-Condon overlaps of the the type $F_{\mu_e \nu_g}^{\mu_g \nu_e} / F_{\mu_g \nu_e}^{\mu_e \nu_g}$ are close to unity and all other close to zero, the upper left part of the adiabatic vibronic Hamiltonian can be approximated by the corresponding adiabatic electronic one. For larger Huang-Rhys factors ($S \geq 0.05$) the mixing with the vibrationally excited vibronic states becomes important.

Bibliography

[1] R. E. Blankenship, *Molecular mechanisms of photosynthesis* (Wiley-Blackwell: Oxford, 2011).

[2] T. Förster, "Zwischenmolekulare Energiewanderung und Fluoreszenz", Ann. Phys. **437**, 55 (1948).

[3] G. McDermott et al., "Crystal structure of an integral membrane light-harvesting complex from photosynthetic bacteria", Nature **374**, 517 (1995).

[4] M. Chachisvilis et al., "Excitons in photosynthetic purple bacteria: Wavelike motion or incoherent hopping?", J. Phys. Chem. B **101**, 7275 (1997).

[5] B. P. Krueger et al., "Calculation of couplings and energy-transfer pathways between the pigments of LH2 by the ab initio transition density cube method", J. Phys. Chem. B **102**, 5378 (1998).

[6] V. Sundström et al., "Photosynthetic light-harvesting: Reconciling dynamics and structure of purple bacterial LH2 reveals function of photosynthetic unit", J. Phys. Chem. B **103**, 2327 (1999).

[7] G. D. Scholes and G. R. Fleming, "On the mechanism of light harvesting in photosynthetic purple bacteria: B800 to B850 energy transfer", J. Phys. Chem. B **104**, 1854 (2000).

[8] J. L. Herek et al., "B800->B850 energy transfer mechanism in bacterial LH2 complexes investigated by B800 pigment exchange", Biophys. J. **78**, 2590 (2000).

[9] M. Wendling et al., "Electron vibrational coupling in the Fenna Matthews Olson complex of Prosthecochloris aestuarii determined by temperature dependent absorption and fluorescence line-narrowing measurements", J. Phys. Chem. B **104**, 5825 (2000).

[10] M Dahlbom et al., "Collective excitation dynamics and polaron formation in molecular aggregates", Chem. Phys. Lett. **364**, 556 (2002).

[11] O. Kühn et al., "Fluorescence depolarization dynamics in the B850 complex of purple bacteria", Chem. Phys. **275**, 15 (2002).

[12] M. Wendling et al., "The quantitative relationship between structure and polarized spectroscopy in the FMO complex of Prosthecochloris aestuarii: Refining experiments and simulations", Photosynth. Res. **71**, 99 (2002).

[13] T. Renger and R. A. Marcus, "On the relation of protein dynamics and exciton relaxation in pigment–protein complexes: An estimation of the spectral density and a theory for the calculation of optical spectra", J. Chem. Phys. **116**, 9997 (2002).

[14] D. Rutkauskas et al., "Fluorescence spectral fluctuations of single LH2 complexes from Rhodopseudomonas acidophila strain 10050", Biochemistry **43**, 4431 (2004).

[15] V. Novoderezhkin et al., "Dynamics of the emission spectrum of a single LH2 complex: Interplay of slow and fast nuclear motions", Biophys. J. **90**, 2890 (2006).

[16] G. S. Engel et al., "Evidence for wavelike energy transfer through quantum coherence in photosynthetic systems", Nature **446**, 782 (2007).

[17] D. Egorova, "Detection of electronic and vibrational coherences in molecular systems by 2D electronic photon echo spectroscopy", Chem. Phys. **347**, 166 (2008).

[18] F. Milota et al., "Two-dimensional electronic photon echoes of a double band J-aggregate: Quantum oscillatory motion versus exciton relaxation", Chem. Phys. **357**, 45 (2009).

[19] A. Nemeth et al., "Double-quantum two-dimensional electronic spectroscopy of a three-level system: Experiments and simulations", J. Chem. Phys. **133**, 094505 (2010).

[20] V. Butkus et al., "Molecular vibrations-induced quantum beats in two-dimensional electronic spectroscopy", J. Chem. Phys. **137**, 044513 (2012).

[21] N. Christensson et al., "Origin of long-lived coherences in light-harvesting complexes", J. Phys. Chem. B **116**, 7449 (2012).

[22] S. Polyutov et al., "Exciton-vibrational coupling in molecular aggregates: Electronic versus vibronic dimer", Chem. Phys. **394**, 21 (2012).

[23] T. Renger et al., "Normal mode analysis of the spectral density of the Fenna-Matthews-Olson light-harvesting protein: How the protein dissipates the excess energy of excitons", J. Phys. Chem. B **116**, 14565 (2012).

[24] V. Tiwari et al., "Electronic resonance with anticorrelated pigment vibrations drives photosynthetic energy transfer outside the adiabatic framework", Proc. Natl. Acad. Sci. USA **110**, 1203 (2013).

[25] L. Valkunas et al., "Vibrational vs. electronic coherences in 2D spectrum of molecular systems", Chem. Phys. Lett. **545**, 40 (2012).

[26] V. Butkus et al., "Distinctive character of electronic and vibrational coherences in disordered molecular aggregates", Chem. Phys. Lett. **587**, 93 (2013).

[27] A. Chenu et al., "Enhancement of vibronic and ground-state vibrational coherences in 2D spectra of photosynthetic complexes", Sci. Rep. **3**, 2029 (2013).

[28] A. W. Chin et al., "The role of non-equilibrium vibrational struc-
tures in electronic coherence and recoherence in pigment–protein
complexes", Nat. Phys. **9**, 113 (2013).

[29] R. Hildner et al., "Quantum coherent energy transfer over vary-
ing pathways in single light-harvesting complexes", Science **340**,
1448 (2013).

[30] P. Nalbach et al., "Vibronic speed-up of the excitation energy transfer in the Fenna-Matthews-Olson complex", http://
arxiv.org/abs/1311.6363 (2013).

[31] V. May and O. Kühn, *Charge and energy transfer dynamics in
molecular systems*, 3rd edition (Wiley-VCH: Weinheim, 2011).

[32] N. Makri, "The linear response approximation and its lowest
order corrections: an influence functional approach", J. Phys.
Chem. B **103**, 2823 (1999).

[33] P. Nalbach et al., "Iterative path-integral algorithm versus cu-
mulant time-nonlocal master equation approach for dissipative biomolecular exciton transport", New J. Phys. **13**, 063040
(2011).

[34] Y. Tanimura and R. Kubo, "Time evolution of a quantum system
in contact with a nearly Gaussian-Markoffian noise bath", J.
Phys. Soc. Japan **58**, 101 (1989).

[35] Q. Shi et al., "Efficient hierarchical Liouville space propagator
to quantum dissipative dynamics", J. Chem. Phys. **130**, 084105
(2009).

[36] K.-B. Zhu et al., "Hierarchical dynamics of correlated system-
environment coherence and optical spectroscopy", J. Phys. Chem.
B **115**, 5678 (2011).

[37] J. Strümpfer and K. Schulten, "Open quantum dynamics calculations with the hierarchy equations of motion on parallel
computers", J. Chem. Theory Comput. **8**, 2808 (2012).

[38] M. H. Beck et al., "The multiconfiguration time-dependent Hartree (MCTDH) method: A highly efficient algorithm for propagating wavepackets", Phys. Rep. **324**, 1 (2000).

[39] H. Wang and M. Thoss, "Multilayer formulation of the multi-configuration time-dependent Hartree theory", J. Chem. Phys. **119**, 1289 (2003).

[40] O. Vendrell and H.-D. Meyer, "Multilayer multiconfiguration time-dependent hartree method: implementation and applications to a henon-heiles hamiltonian and to pyrazine", J. Chem. Phys. **134**, 044135 (2011).

[41] J. Schulze, "Highdimensional exciton dynamics in photosynthetic complexes", Master Thesis (University of Rostock, 2014).

[42] M. E. Madjet et al., "Intermolecular Coulomb couplings from ab initio electrostatic potentials: Application to optical transitions of strongly coupled pigments in photosynthetic antennae and reaction centers", J. Phys. Chem. B **110**, 17268 (2006).

[43] P.-A. Plötz et al., "A new efficient method for calculation of Frenkel exciton parameters in molecular aggregates", J. Chem. Phys. **140**, 174101 (2014).

[44] A. Caldeira and A. Leggett, "Path integral approach to quantum Brownian motion", Phys. A **121**, 587 (1983).

[45] A. Ishizaki and Y. Tanimura, "Modeling vibrational dephasing and energy relaxation of intramolecular anharmonic modes for multidimensional infrared spectroscopies", J. Chem. Phys. **125**, 084501 (2006).

[46] S. Mukamel, *Principles of nonlinear spectroscopy* (Oxford University Press: New York, 1995).

[47] T. Renger and V May, "Ultrafast exciton motion in photosynthetic antenna systems: The FMO-complex", J. Phys. Chem. A **102**, 4381 (1998).

[48] J. Adolphs and T. Renger, "How proteins trigger excitation energy transfer in the FMO complex of green sulfur bacteria", Biophys. J. **91**, 2778 (2006).

[49] A. Ishizaki and G. R. Fleming, "On the adequacy of the Redfield equation and related approaches to the study of quantum dynamics in electronic energy transfer", J. Chem. Phys. **130**, 234110 (2009).

[50] Y. Tanimura, "Stochastic Liouville, Langevin, Fokker-Planck, and Master Equation approaches to quantum dissipative systems", J. Phys. Soc. Japan **75**, 082001 (2006).

[51] R.-X. Xu et al., "Exact quantum master equation via the calculus on path integrals", J. Chem. Phys. **122**, 41103 (2005).

[52] R.-X. Xu and Y. Yan, "Dynamics of quantum dissipation systems interacting with bosonic canonical bath: Hierarchical equations of motion approach", Phys. Rev. E. **75**, 031107 (2007).

[53] J. Shao, "Decoupling quantum dissipation interaction via stochastic fields", J. Chem. Phys. **120**, 5053 (2004).

[54] Y.-a. Yan et al., "Hierarchical approach based on stochastic decoupling to dissipative systems", Chem. Phys. Lett. **395**, 216 (2004).

[55] R. Feynman and F. Vernon, "The theory of a general quantum system interacting with a linear dissipative system", Ann. Phys. **24**, 118 (1963).

[56] Y. Tanimura and A. Ishizaki, "Quantum dynamics of system strongly coupled to low-temperature colored noise bath: Reduced hierarchy equations approach", J. Phys. Soc. Japan **74**, 3131 (2005).

[57] Y. Jing et al., "Equilibrium excited state and emission spectra of molecular aggregates from the hierarchical equations of motion approach", J. Chem. Phys. **138**, 045101 (2013).

[58] C. Kreisbeck et al., "High-performance solution of hierarchical equations of motion for studying energy transfer in light-harvesting complexes", J. Chem. Theory Comput. **7**, 2166 (2011).

[59] C. Olbrich et al., "Theory and simulation of the environmental effects on FMO electronic transitions", J. Phys. Chem. Lett. **2011**, 1771 (2011).

[60] U. Weiss, *Quantum dissipative systems*, Vol. 13, Series in Modern Condensed Matter Physics (World Scientific Publishing: Singapore, 2008).

[61] C. Meier and D. J. Tannor, "Non-Markovian evolution of the density operator in the presence of strong laser fields", J. Chem. Phys. **111**, 3365 (1999).

[62] B. Hein et al., "Modelling of oscillations in two-dimensional echo-spectra of the Fenna–Matthews–Olson complex", New J. Phys. **14**, 023018 (2012).

[63] R. Berera et al., "Ultrafast transient absorption spectroscopy: Principles and application to photosynthetic systems", Photosynth. Res. **101**, 105 (2009).

[64] D. M. Jonas, "Two-dimensional femtosecond spectroscopy", Annu. Rev. Phys. Chem. **54**, 425 (2003).

[65] M. Cho, "Coherent two-dimensional optical spectroscopy", Chem. Rev. **108**, 1331 (2008).

[66] T. Brixner et al., "Two-dimensional spectroscopy of electronic couplings in photosynthesis", Nature **434**, 625 (2005).

[67] J. Dostál et al., "Two-dimensional electronic spectroscopy reveals ultrafast energy diffusion in chlorosomes", J. Am. Chem. Soc. **134**, 11611 (2012).

[68] P. Hamm and M. Zanni, *Concepts and methods of 2D infrared spectroscopy* (Cambridge University Press: Cambridge, 2011).

[69] M. Khalil et al., "Obtaining absorptive line shapes in two-dimensional infrared vibrational correlation spectra", Phys. Rev. Lett. **90**, 047401 (2003).

[70] A. Ishizaki and Y. Tanimura, "Dynamics of a multimode system coupled to multiple heat baths probed by two-dimensional infrared spectroscopy", J. Phys. Chem. A **111**, 9269 (2007).

[71] L. Chen et al., "Two-dimensional electronic spectra from the hierarchical equations of motion method: Application to model dimers", J. Chem. Phys. **132**, 024505 (2010).

[72] A. Halpin et al., "Two-dimensional spectroscopy of a molecular dimer unveils the effects of vibronic coupling on exciton coherences", Nat. Chem. **99**, 1 (2014).

[73] J. Yuen-Zhou et al., "A witness for coherent electronic vs vibronic-only oscillations in ultrafast spectroscopy", J. Chem. Phys. **136**, 234501 (2012).

[74] L. Chen et al., "Optical line shapes of molecular aggregates: Hierarchical equations of motion method", J. Chem. Phys. **131**, 094502 (2009).

[75] C. C. Jumper et al., "Intramolecular radiationless transitions dominate exciton relaxation dynamics", Chem. Phys. Lett. **599**, 23 (2014).

[76] M. Schröter and O. Kühn, "Interplay between nonadiabatic dynamics and Frenkel exciton transfer in molecular aggregates: Formulation and application to a perylene bismide model", J. Phys. Chem. A **117**, 7580 (2013).

[77] J. Schulze et al., "Exciton coupling induces vibronic hyperchromism in light-harvest- ing complexes", New J. Phys. **16**, 045010 (2014).

[78] E. Bašinskaite et al., "Vibronic models for nonlinear spectroscopy simulations", Photosynth. Res. **121**, 95 (2014).

[79] M. R. Philpott, "Some modern aspects of exciton theory", Adv. Chem. Phys. **23**, 227 (1973).

[80] F. C. Spano, "Absorption and emission in oligo-phenylene vinylene nanoaggregates: The role of disorder and structural defects", J. Chem. Phys. **116**, 5877 (2002).

[81] H. Marciniak et al., "One-dimensional exciton diffusion in perylene bisimide aggregates", J. Phys. Chem. A **115**, 648 (2011).

[82] D Ambrosek et al., "Quantum chemical parametrization and spectroscopic characterization of the Frenkel exciton Hamiltonian for a J-aggregate forming perylene bisimide dye", J. Phys. Chem. A **116**, 11451 (2012).

[83] F. Fennel et al., "Biphasic self-assembly pathways and size-dependent photophysical properties of perylene bisimide dye aggregates", J. Am. Chem. Soc. **135**, 18722 (2013).

[84] M. T. W. Milder et al., "Revisiting the optical properties of the FMO protein", Photosynth. Res. **104**, 257 (2010).

[85] D. E. Tronrud et al., "The structural basis for the difference in absorbance spectra for the FMO antenna protein from various green sulfur bacteria", Photosynth. Res. **100**, 79 (2009).

[86] H. M. Berman et al., "The Protein Data Bank", Nucleic Acids Res. **28**, 235 (2000).

[87] W. Humphrey et al., "VMD – Visual Molecular Dynamics", Journal of Molecular Graphics **14**, 33 (1996).

[88] M. Schmidt am Busch et al., "The eighth bacteriochlorophyll completes the excitation energy funnel in the FMO protein", J. Phys. Chem. Lett. **2**, 93 (2011).

[89] J. Moix et al., "Efficient energy transfer in light-harvesting systems, III: the influence of the eighth bacteriochlorophyll on the dynamics and efficiency in FMO", J. Phys. Chem. Lett. **2**, 3045 (2011).

[90] A. Ishizaki and G. R. Fleming, "Theoretical examination of quantum coherence in a photosynthetic system at physiological temperature", Proc. Natl. Acad. Sci. U. S. A. **106**, 17255 (2009).